Are Quanta Real?

Are Quanta Real?

A GALILEAN

DIALOGUE

J. M. JAUCH

Josef

Indiana University Press

BLOOMINGTON / LONDON

Contents

Preface of the Reporter

THE CONVERSATION RECORDED IN THIS DIALOGUE WAS in the final stages of editing when I had the unexpected pleasure of meeting Giovanfrancesco Sagredo on one of my early morning walks along the shores of Lake Geneva. He sat in a pensive mood on a bench as if surrounded by a halo of refined culture and elegance. When he saw me he greeted me with warmth as though he had expected me, saying: "Your latest version of our dialogue is much improved over the first. I am very pleased with it now. But I am not quite sure whether it is really a true rendering of our actual discussion. You must have put a lot of work and some of your own ideas into the text."

Uncertain whether this was a slight criticism in the form of a compliment, I said to him: "The credit for whatever improvement was made should not go to me but to the many friends and colleagues who contributed valuable criticisms and comments. I cannot possibly mention them all; there are too many. But among them are some whose contributions have been particularly extensive, as, for example, those of Prof. F.J. Belinfante of Purdue University, Prof. C.A. Hooker of the University of Western Ontario, Canada, Prof. R. Kromhout and Prof. J.B. Dence of Florida State

University, Prof. E.P. Wigner of Princeton University, Prof. R.G. Newton of Indiana University, Prof. F. Rohrlich of Syracuse University, and Dr. G. Baron of Rye, New York. Above all I should mention the work of my friend and colleague Constantin Piron, who influenced all my thoughts on this topic."

Sagredo smiled and said: "And your charming secretary did not mind retyping the several versions of the manuscript?" "Not only did she not mind but she did it with such a cheerful disposition that it was a real pleasure to work with her," I said. "Finally," I added, "you should not forget my editor, who went over the entire text with that patience and sensitivity which are the hallmark of true devotion."

Sagredo, with an incomparably graceful smile, replied warmly: "So none of us can lay exclusive claim to the final product. We see here an example of our conclusion of the fourth and final day of our discussion: *That the whole is more than the sum of its parts and that the constructive interplay of complementary processes is the secret of all creative activity in life.*"

J.M. JAUCH

Geneva, July 1971

Introduction

THE CENTRAL QUESTION AT THE SOLVAY CONGRESS IN 1927 was: Is quantum mechanics a "complete theory," or is the statistical character of its predictions merely the reflection of our ignorance of an underlying causal infrastructure? Today, after forty years of unchecked success of quantum mechanics on every front of physical science, the question still occupies the minds of many thoughtful students of quantum mechanics.

The majority of the participants at the Solvay Congress were of the opinion that the probabilistic interpretation of the state vector was not merely a reflection of our ignorance but rather the essential physical content of Planck's quantum of action. Yet an important minority, among them Einstein, Schrödinger, and de Broglie, held the opposite view.

The question cannot of course be decided unless one can construct a causal theory which permits a confrontation with experimental facts. No one has succeeded in doing so. It is, however, possible to speculate on what such causal infrastructures would have to look like if the theory is not to lead to predictions which disagree with known facts.

These are the theories with "hidden variables." Numerous attempts have been made to construct such theories.

Before one starts to examine in detail such speculative theories, it is useful to recall some general aspects of the problem.

It is rarely sufficiently emphasized that the classical concept of causality (which in its physical context is more aptly called determinism) is in fact a gigantic prejudice, which is often wrongly identified with the very essence of science. Yet, there were occasional, isolated voices which drew attention to the flimsy evidence for such a point of view, and which steadfastly maintained that there was in fact more evidence in favor of statistical laws in physics.

One of the most remarkable spokesmen for this point of view was the American philosopher Charles S. Peirce, who devoted most of his life to the analysis of the basic logic and structure of the physical sciences.

Peirce traced the origin of the idea of necessity to Democritos and opposed it with the ideas of Epicurus. He concluded that the only natural laws which can be validly inferred from observations are laws of chance. His pertinent observations are even more relevant today than they were in his time and merit attention by all who are interested in these questions.

Next, it must be emphasized that the inferences which are drawn from observations and then incorporated into a theory are nondeductive. Most are of three kinds : induction, hypothesis, or analogy. All have their roots in extra-scientific domains such as habits, traditions, esthetic visions, or ideologies.

The quest for hidden variables in quantum mechanics has its roots in a past ideology, *viz.*, the determinism of nineteenth-century materialism. Thus, far from being a vision of a future theory, it harks back to a glorious past, but nevertheless a past which seems to recede rapidly into the distance under the impact of new evidence, giving way to new forms of scientific thought.

Seen in this perspective, it is not surprising that the discussions which surround the quest for hidden variables in quantum mechanics have, on both sides of the camp, often been conducted in a spirit of aggressiveness which resembles more the defense of orthodoxy of one ideology or another than a spirit of scientific objectivity.

Yet, there are scientific aspects of the problem which are extremely interesting and which are worthy of thorough exploration. Their study not only gives us a better understanding of the epistemological problems of quantum mechanics but they may also lead to generalizations or modifications of this theory which could be indispensable for future progress in microphysics.

The situation presents a remarkable similarity with that which occurred at the beginning of the seventeenth century, when the geocentric system of Ptolemaeus had to give way to the heliocentric one of Copernicus.

Then as now the question could not be decided on empirical grounds alone since both systems were capable of correctly describing the observed phenomena. Then as now the debate was strongly motivated by ideological considerations, and then as now the new view was often opposed by an appeal to reasons which Galileo showed were completely invalid. It is this last point which inspired Galileo in 1630 to write his celebrated "Dialogue on the Two Major Systems of the World."

The close similarity of the general epistemological situation has induced me to appeal to the three imaginary interlocutors who carried on the famous "dialogue" during a critical moment in the history of science to give us the benefit of their wisdom at a juncture of history which is, perhaps, comparable in importance to that of three hundred years ago. The dialogue form was found to be an ideal way of reproducing the dialectical process of arriving at a deeper understanding of the enigmas which quantum mechanics presents us.

Most of the puzzling features in this new epistemology can be understood without technical knowledge. Some basic facts suffice. Many of the new insights gained by the analysis of these questions have wide repercussions and provide new perspectives, which transcend all levels of human activity.

With a few exceptions destined for the specialist, all of this dialogue can be read and understood by anybody with sufficient interest in the epistemology of modern science.

Many of the passages used are more or less faithful reproductions of actual conversations or statements from correspondence and published material. The three interlocutors do not represent actual persons, however. They are composite characters, each representing a current tendency.

I hope that living persons who find themselves thus "quoted" are satisfied as to the accurateness with which their opinions are represented.

Dialogue on the Question

Are Quanta Real?

BETWEEN

Filippo Salviati, Giovanfrancesco Sagredo,

and Simplicio

TIME AND PLACE: *Fall of 1970, in a villa*

on the shores of Lake Geneva.

FIRST DAY

SALVIATI It was our resolution at the end of the fourth day of that memorable year, 1638, that we should reconvene our assembly when it would be less inconvenient for us, to satisfy our desire concerning the problems that remain to be discussed.

I propose that we should discourse as distinctly and concretely as possible on the natural reasons hitherto alleged on one side by those who maintain the deterministic, materialistic philosophy, and on the other by the followers of the Copenhagen interpretation of quantum mechanics.

Bohr, by denying objective reality to properties whose simultaneous presence would require for their verification mutually exclusive physical situations, puts in question that very concept of reality which has been the cornerstone of all of physics as it has developed from the time of our last discussions until the present.

It would therefore be good if we began our disputation with the examination of what and how great is the strength of the Materialists' argument when they claim that Bohr's hypothesis of complementarity is unacceptable to them.

They say that the world has a real existence independent of our observation; that our science should reveal the reality of this world and determine its laws, which are absolute and immutable.

They give many reasons for this view which are familiar to you, not least of which is the fact that we all live in accordance with this view in practice, and it is therefore confirmed by common sense.

SIMPLICIO   Not only is it common sense, but the reality of the world can be proved both logically with the most subtle of reasons and by experimentation.

SAGREDO   My most respected friends, let us not prejudge our case. As for me, I am not learned enough to follow these most subtle reasons which you cite in support of the thesis of the reality of the world, and as for the experiments that you mention, I must admit that I have never heard of them.

You two seemed to be much more versed in these matters than I could ever hope to be; nevertheless I venture to guess from our previous conversations that you do not hold the same views as to the question concerning the reality of the world and the objects that we find therein. So please do not expect from me any definite view on the matter before I have heard your arguments for or against the points in dispute.

SALVIATI   Well said, my dear friend, and thank you for reminding us that we should advance more cautiously and state our views before we allege proofs on matters which we should scrutinize with the utmost care and openness of mind.

Therefore, let us begin by considering one of the simplest possible situations which would reveal the complementarity of certain physical properties.

I have brought with me two pieces of polaroid, a material which you all know and have certainly used on many occasions.

SIMPLICIO   Yes, I know it very well since I use sunglasses made of polaroid.

SALVIATI   Then you will be familiar with some of their properties. For example, when we look at the sky with them we find that parts of the sky are darkened or brightened, depending on the orientation of the polaroid with respect to the line of vision.

SIMPLICIO   We have all seen these things from our school days; and everybody knows that the transverse character of wave propagation, as it was explained by Fresnel in his celebrated memoir of 1822, is responsible for this.[1]

SAGREDO   I believe that the remarkable phenomena to which Salviati refers actually show more than that. They show us also that the light of the sky is partly polarized, so that it has lost full symmetry around the line of sight when it reaches our eye, and that the polaroid reveals this asymmetry.

SALVIATI   Very well said indeed! It is this acute intelligence of yours which makes this discussion such a pleasure and justifies my hope of eventually reaching full understanding in the difficult matters that we face.

What we see here might be reproduced in a more striking manner by using not one but two polaroid plates, one behind the other. Looking through such a pair at any source of light, we find that the intensity depends on the relative orientation of the two around the line of vision.

There is one orientation for which the intensity of transmitted light is maximal. Starting from this position I turn the second of the polaroid sheets around the axis and you can see that the intensity of the transmitted light diminishes. It

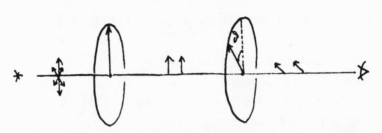

FIG. I. *Observing a source of light through two sheets of polaroid.*

reaches a minimum of complete darkness when the second polaroid has been rotated 90° with respect to the first. If I turn beyond 90°, the brightness returns, until at 180° it reaches its maximal value again.

How can we understand this remarkable behavior of light when it traverses our sheets of polaroid?

SIMPLICIO   These things are now very well understood, since Fresnel taught that light is a transverse vibration, and Maxwell later showed that the vibration is that of an electromagnetic field which propagates through space with the speed of light.

SAGREDO   Do I understand you to say that light is a wave motion, propagating through space somewhat as water waves propagate over the surface of water and sound waves propagate through the atmosphere?

SIMPLICIO   Exactly. The only difference is that electromagnetic waves propagate through *empty* spaces and they are transverse waves, as is revealed by Salviati's experiments with polaroid.[2]

SAGREDO   I notice with astonishment that you have turned away completely from the teachings of Aristotle, which you once defended with such vigor.

You know of course that Aristotle denied the existence of the void, and that his space is filled with a substance. How much easier it would be to have this substance at our disposal now as the carrier of light waves, just as water carries water waves and air carries sound waves. Why do you refuse to admit a substance as the carrier of light waves?[3]

SIMPLICIO   While it is true that I once considered Aristotle the greatest philosopher who ever lived, you cannot reproach me for having actively defended his thesis of the plenum of space. I have long since recognized that Democritos came much closer to the truth in these matters, and that a void populated by real atoms and penetrated by fields of various kinds is the correct representation of our physical universe.

SAGREDO   If I understand you correctly, you have two kinds of realities in your physical universe: atoms and fields.[4]

SIMPLICIO   That is correct. And both these realities are subject to a dynamical law of evolution which is partly known to us today and which we shall eventually discover completely, thus giving us a full understanding of basic physical structures.

SAGREDO   But, pray, you have not yet explained to us why you deny the existence of a substance as a carrier of light waves.

SIMPLICIO   I am not a physicist, as you well know, but I have been told by my physicist friends that the hypothesis of a substance as a carrier of light waves is not only unnecessary, but leads to difficulties when the propagation of light is studied with respect to different observers moving with respect to one another.

SALVIATI   Let us not stray too far from our objective, because we have a long way to go yet and the road is hard. We are fortunate to be able to enjoy Sagredo's stimulating and penetrating questions and to profit from Simplicio's erudition. But his answers, although seemingly perfect in their clarity and precision, will have to face the test of the reality of the phenomena, which are after all our last judges for whatever theoretical representation we might propose.

Let us then reconsider our experiment with the two sheets of polaroid, and let us now diminish the intensity of the light. What will happen to our polarization experiment?

SIMPLICIO The intensity of light transmitted through the two polarizers decreases in equal proportion for all relative positions of the two plates.

SALVIATI Right you are, and indeed it is easy to observe this. However, I am now concerned with the question of whether this observation remains true down to the smallest possible intensities.

SIMPLICIO I do not see the point of your question, unless you mean to ask me what happens when the corpuscular nature of light is brought into evidence.

SALVIATI This is exactly what I am driving at, and since you have already anticipated my question let me be more precise, and take for the source of light one single atom, which passes from an excited state of energy E_1 to a ground state of energy E_0, thereby emitting a single quantum of light, a photon, of frequency

$$\nu = \frac{E_1 - E_0}{h}$$

(where h is Planck's constant).

This photon will fall on the two plates of polaroid, and I now ask the question, will it go through both or will it be absorbed in the second plate?[5]

SAGREDO I find this a most baffling question indeed. Although I am not always sure whether our friend Simplicio is correct in his answers, the history of science since our last discussion in 1630 has shown that he was monumentally wrong on many points. Yet, I cannot deny that I rather liked his explanation with the transverse polarized electromagnetic waves. What troubles me is that I cannot possibly find in this picture an answer to your question, Salviati.[6]

SALVIATI   It is indeed a difficult question, and I would go even further than you, Sagredo, because I am firmly convinced that it is not only difficult to find an answer but impossible to do so with Simplicio's theory of the transverse wave.

SAGREDO   What do you mean? The theory of the propagation of light as a transverse wave is supported by hundreds of the most refined experiments, from Young and Fresnel down to the present. How can such a powerful, efficient, and beautiful theory be wrong? You question one hundred and fifty years of physical optics!

SALVIATI   I do not imply that the experiments on physical optics to which you allude are wrong. I mean to say that wave theory, which Simplicio has so ably explained to us, cannot possibly predict what happens to a *single photon* when it traverses two crossed polaroids. If you do not believe my statement, try to do it. I shall be the first to acknowledge defeat if you suceed.

None of your physicists who developed physical optics worked with single photons, so none of them could possibly have discovered the breakdown of the theory when it is applied to an individual photon.

Indeed, the very concept of a photon as a sort of atom of light is entirely alien to the conceptual structure of a field, such as an electromagnetic wave. The former is a discrete, indivisible entity, the latter a continuum. These two concepts are hard to reconcile in one and the same physical object.

SAGREDO   I am profoundly disturbed about the disruption of the unity of science which you indicate with your remarks. It is as though there were a basic duality in nature which furnishes complementary descriptions of one and the same object.

For, if I answer your question with the statement that the photon will penetrate the second polaroid sheet, then of course I see no reason why there should be any darkening at all when the relative position of the polaroids is between 0° and 90°. If on the other hand I answer the question by saying that the photon *will not* penetrate, then there should be complete darkness behind the second polaroid. Since neither one nor the other is observed, neither of my answers can be correct.

SALVIATI And indeed neither is correct, or rather both are correct and both are wrong.

SIMPLICIO You are surely not serious with your last statement!

SALVIATI Indeed I am, and I can prove the statement too!

SIMPLICIO Take care, Salviati; you are very ambitious. How can you hope to prove something which is manifestly against all sound reason? For either the photon does or does not penetrate the two polarizers, and it seems to me there is no third position possible between these incompatible opposites.

SALVIATI How limited is man's reason before the mystery of nature! We have reached the limit of reason and we must go humbly to nature and ask her the question. She will answer and her answer is truer than any answer that you or I or any one of us might conceive.

So for this purpose I have arranged a little experiment: The source of light is replaced by a single atom, which ultimately will emit a photon by passing from an excited state of energy E_1 to a ground state E_0. Once the emission is completed, I replace it by a second, similar atom, and then by a third, and so on. In place of the eye I put an instrument called a photomultiplier tube, whose sole purpose is to register the arrival of the photon behind the second polarizer, whose axis makes an angle of $\vartheta = 45°$ with respect to the first. The whole experiment is mechanized and produces at the end a strip of paper with a string of o's and 1's printed on it. o signifies that the photon did not arrive, and 1 signifies that it did arrive at the photomultiplier tube. I set the experiment in motion by turning a switch, and now let us see what happens.

There, the strip is coming out and we find it looks as follows:

1 1 0 1 0 1 1 0 1 0 1 0 1 0 0 1 1 0

SIMPLICIO   Something must be wrong with your machine, because it does not seem to give a definite answer to our question.

SALVIATI   But, Simplicio, you did not expect a definite answer, since you have proved that no definite answer is possible, as either interpretation leads to a contradiction with the phenomena observed at a high intensity of light.

SIMPLICIO   That is true, but I did not expect this answer either. Perhaps there is a regularity in this sequence of zeros and ones which escapes us. Certainly it cannot be a simple alternation, and I cannot detect a more hidden regularity without more evidence.

SALVIATI   I shall let the machine operate while we continue our discussion, and then you will be at liberty to examine the results. In fact, to make the work easier for you

I shall store the information in this large computer, which I connect to the output of the photomultiplier. The computer is programmed to carry out a number of tests for randomness as the series develops, so any possible regularity that we can imagine will eventually reveal itself.

SAGREDO This sequence of zeros and ones which you showed us so far is just like the sequence of heads and tails that one observes in tossing a coin!

SALVIATI Quite true, and in fact I can reveal that I have run this test many times and I have never been able to detect any essential difference between the observed sequences and a random sequence which would be observed in the manner you describe.

SAGREDO Should we not conclude from this that the same, or perhaps a similar, mechanism that produces the random sequence in the coin-tossing experiment is responsible for the random sequence in your polarization experiment?

SIMPLICIO Of course, we all know that different effects must proceed from different causes. This is a principle as fundamental as the principle of contradiction. In the case of the coin-tossing we produce the random sequence by the inevitable slight variations of the initial conditions and the slight variations of the physical conditions during the time of the throw. If we were able to control these conditions with sufficient accuracy, then the randomness of the sequence would give way to a perfectly regular sequence of either only heads or only tails. We are sure of this because we know that the equations of motion are deterministic and that they are consistent with this affirmation. Furthermore, I actually knew a man who had such a fine hand and was so skillful that he could almost always throw a head or a tail as he desired. I have tried it myself, and with a little practice I could produce one of the desired results with about 80%

probability, enough to make a handsome profit in a betting game.

Thus, I do not doubt that each individual photon which is emitted from the source is physically slightly different from any other, a difference which will eventually reveal itself by either the transmission or the absorption in the second polaroid plate.

There is therefore nothing mysterious about the fact that we observe a random sequence in the photon experiment. It merely reveals that a *hidden variable* is needed for a full description of the physical state of the photon, and that this hidden variable has different values for each of the individual photons we observe.

Salviati   When you say *hidden*, do you mean to say that this variable is inaccessible to further determination or measurement?

Simplicio   You are trying to trap me. I know very well the difficulties of giving either answer to your question. I said, of course, *hidden* merely because, just as in the coin-tossing experiment, it would be extremely difficult to fix the conditions of the light source with such precision that the outcome of all future physical measurements would be completely determined by these conditions. But you would certainly not deny that the physical conditions of the photon-emitting atoms cannot be completely identical, hence that there is certainly room for a more refined variety of initial conditions which could be correlated to the outcome of the experiment?

Salviati   I am afraid, Simplicio, that you are overlooking an important point in your analogy of the photon experiment with the coin-tossing experiment.

Simplicio   What could that be?

Salviati   In the photon experiment we use two polarizing plates. It is only the second plate which corresponds to the revelation of the fact of whether the coin shows head or tail.

SIMPLICIO   This is no serious problem, since we can include the first plate with that part of the experiment which prepares the state, and then the analogy with the coin-tossing is restored.

SALVIATI   Very well, but now please tell me, do you attribute the distribution of the outcome at the second plate to the source or to the first plate?

SIMPLICIO   I would say to the source, since the first plate is a simple system and always remains the same while the source changes from one photon to the next.

SALVIATI   This seems to be a sensible answer. Let us see whether it can be maintained. In order to test it I have here a third plate, which I place behind the other two, and I adjust its axis of polarization so that it makes an angle of 45° with the axis of the second and an angle of 90° with the axis of the first. Let us see what we observe. Here it comes. Again we find a sequence of zeros and ones, but the sequence looks a bit different:

0 1 0 0 0 1 0 0 0 0 0 1 0 0 0 0 0 1 0 1 0 0 0 1 0 1 0 0 0 1 0 0.

You notice no doubt that in this sequence we have many fewer ones than zeros; in fact, there are only 8 ones and 24 zeros.

SIMPLICIO   Yes, I see that.

SAGREDO   It is just like flipping coins when the winning game is two heads in a row. I think that would give the same distribution.

SALVIATI   That is perfectly true. But let us not forget that the purpose of the experiment was to test whether the hidden variables are associated with the source of light, or whether we should hold the polarizing plates responsible for their values.

SIMPLICIO   How can we test that with this experiment?

SALVIATI   Very easily. By a slight variation of the preceding experiment, which incidentally keeps on going and has already produced a sequence of hundreds of digits while we were talking.

You can see that the ones keep on showing up, but only about one-fourth of the time.

Let us now change nothing at all on the source and remove only the middle plate. We are now left with only two polarizing plates, whose axes of polarization are at 90° with respect to one another. I remind you of the arrangement with a little sketch.

Here is the result of the experiment:

o o o o o o o o o o o o o o o o o o o o o o o o

It is a perfect sequence of zeros. The arrangement is completely opaque, quite contrary to the preceding experiments.

Let us keep in mind that I have removed only the middle plate. I have not touched the source or the other plates or the

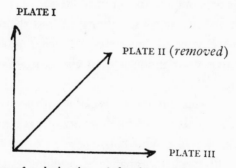

FIG. 2. *Axes of polarization of the three plates.*

detector. Simplicio, would you not agree that the middle plate has a decisive influence on your hidden variables, for which you held the source alone responsible?

SIMPLICIO   I must confess I have never made a similar observation. But while it certainly shows an effect by the middle plate on the hidden variables, it is not conclusive evidence that the source has nothing to do with the values of these parameters.

SALVIATI   Correct, nor does it disprove that the phase of the moon, which constellation the sun is in, or the state of my consciousness has nothing to do with the values of these parameters. It leaves the door wide open for all kinds of theories concerning the origin of your hidden variables.[7]

But I see that our friend Sagredo is eager to say something, so let him have a chance to comment on the conclusion we have reached.

SAGREDO   Yes, my dear Salviati. Your beautiful demonstration that the polarizing plates must be responsible for at least some of the values of the hidden parameters reminded me of something quite outside the domain of science, but nevertheless related to our situation by analogy.

Those of us who have the great fortune to live in a nation based on law are used to the idea that an individual accused of a crime is presumed innocent until he is proven guilty.

I have heard that in some nations it is just the other way around. There you are suspected of being guilty unless you can prove your innocence to the satisfaction of the prosecution.

In a way, the second version is easier for the prosecution, just as Simplicio's theory of hidden variables is easier for the physicist. But anyone who knows the first system would agree that it offers a better chance of reaching a just verdict.

Simplicio seems to have chosen a solution for the occurrence of the random sequence in Salviati's photon experiment which corresponds to the second one of this system. He seems to act like a somewhat hard-pressed prosecutor since he finds a random sequence as inacceptable as a crime, and a culprit must be found to atone for it. He has chosen the most obvious one, the source of light. But your experiment with the three polarizing plates showed that the crime is more like a conspiracy. At least we have solid evidence that the middle plate is certainly responsible for some of the crime. And since the other plates are identical with the middle plate in their physical structure, there is good reason to suspect the other plates as well. But when we have so many culprits in the dock the situation becomes bewildering.

Perhaps we have uncovered only a small part of a profound, all-pervading conspiracy to annoy us with random sequences at some crucial points in our quest for understanding.

SIMPLICIO   I do not think that your analogy is fair or even appropriate for my feeling on the matter of random sequences. All I really say is that the photon that penetrates the plates and the photon that gets absorbed must have different real physical properties. And a theoretical description of nature which has no counterpart in its concepts for this difference seems to me an incomplete theory.

In the case of the flipped coin, we found such a counterpart in the initial conditions. In the case of Brownian motion, which was equally baffling at first, we found it in the atomic constitution of matter, a concept which has been of enormous importance, and whose scientific value nobody would question today .

With the photon experiment we have again uncovered a random process not so different from that of Brownian motion, so it seems to me quite natural and in keeping with sound scientific tradition to look for an explanation for the results of your random photon experiment similar to that of Brownian motion.

SALVIATI   You have a way of arguing which is as convincing as it is deceptive. Indeed you place great value on the *similarity* of the situation between the photon experiment and Brownian motion but you choose to ignore the difference. On the basis of this similarity you argue by an analogy which may convince the vulgar.

However, we should not be led to conclusions by majority opinions, because in science democracy is singularly ineffective. The bold and courageous ideas of the nonconformists have always represented the milestones of progress in science. You remember too well our last discussion concerning the opinions of the celebrated Academician about the origin of the tides. Today we know how far he was from the truth, but had he not had the courage to defend a point of view, which was new, original, and fruitful for the future progress of science, we would still be arguing today whether the moon was made of a glassy, perfect, fifth substance.

SIMPLICIO   You speak of differences which I choose to ignore. What are these differences? Let me know them so that I can see for myself whether they are not merely differences in degree rather than differences in essence.

SAGREDO   Yes, Salviati, please tell us which differences force us to adopt a solution for the occurrence of random sequences in your experiment other than the one of hidden variables? I must confess that I have often heard this point discussed. It seems to me the crucial one in the entire logical and phenomenological structure of microphysics. Bohr called this other solution the *principle of complementarity*, and it is possibly one of the greatest discoveries in the scientific history of mankind, with ramifications on many other levels of science.

But I am afraid that I am talking too much. I will defer to Salviati, who is much more versed in these matters. It will be a pleasure to hear him explain it once more with the lucidity and depth of comprehension that we have come to appreciate in him.

Salviati You do me too much honor with your flattering comments about my lucidity and what not. You know very well that I have never hidden my profound humility when I speak about complementarity. It seems to me rather that we have reached a sort of limit in our understanding which is like any other of the contingencies with which we are forced to live. Why do we have two hands and two feet? Why do our eyes see only a tiny fraction of the spectrum? Why is our earth finite and structured as it is? And why should Planck's constant have exactly the value that it has?

That is where it all begins, that singular history of the quantum of action. It is a history of surprises, errors, confusion, and daring vision; it is truly a scientific mutation.[8]

But let me not remain in generalities. You, Simplicio, are impatient to know what it is that is so essentially different in the occurrence of random sequences that could not be rendered by a physical picture analogous to the one we used in the description of Brownian motion.

I am sure you are already aware of these differences. Only in your eagerness to find an explanation of our experiment which conforms to your ideology, you have conveniently forgotten them. So if you permit me, let me ask you a few questions just to refresh your memory.

Simplicio Do not hesitate. I always find your questions challenging.

SALVIATI   You remember that we started our experiment as a simple observation with polaroid filters which any child can do.

SIMPLICIO   Yes, I do remember.

SALVIATI   And that we had no feeling of any particular difficulty in explaining the experiment on the basis of the classical wave theory of light.

SIMPLICIO   There is no difficulty whatsoever. In fact the classical theory yields the correct variation of the intensity as a function of the angle ϑ between the polarization directions of the two plates:

$$I = I_0 \cos^2 \vartheta$$

SALVIATI   You would say then that the difficulty appeared only when we started to perform experiments with individual photons.

SIMPLICIO   Assuredly that is when the difficulty appeared.

SALVIATI   Thus the existence of photons is the source of our problems.

SIMPLICIO   It really seems that way to me.

SALVIATI   Tell me, Simplicio, should we not call the photons the "atoms" of light?

SIMPLICIO   If you mean by "atom" the indivisible in the sense the Greeks meant it, then this designation seems appropriate to me.

SALVIATI   A little while ago you mentioned Brownian motion as an analogy to the occurrence of randomness in physical systems of small dimensions. How did Brownian motion become integrated into the rest of physics?

SIMPLICIO   We call Brownian motion the random fluctuation in position and velocity of very small physical objects which are exposed to the irregular impulses of individual atoms and molecules. When this was recognized one could not only explain the occurrence of this motion but in addition one could verify quantitatively the size and the number of atoms which cause this irregular motion.

SALVIATI   You would therefore say that the hypothesis of the atomic nature of matter *explains* the occurrence of random fluctuations in Brownian motion.

SIMPLICIO   Very truly this is so.

SALVIATI   While with the photons and their random fluctuation the situation is exactly the reverse?

SIMPLICIO   What do you mean?

SALVIATI   Did you not admit just a few minutes ago that it was the "atomic" nature of light which causes all the difficulty in the polarization experiment?

SIMPLICIO   Yes, indeed.

SALVIATI   And did you not explain to us that the theory, which disregards the individuality of photons, gives the correct expression for the variation of the transmitted intensity?

SIMPLICIO   Yes, I did.

SALVIATI   Do you see now, how weak your reasoning by analogy turns out to be when you begin to examine it more carefully?

SIMPLICIO   I do not know what to say; you have succeeded in confusing me, but this shows only that you are cleverer than I in arguing about such matters. I need more time to think about this question.

SALVIATI   Please take all the time you need. I do not really enjoy confusing you. In a sense we are all confused when we have to face the baffling behavior of light and other microobjects. All I am trying to do is to eliminate as many false routes as possible, so that we are left with fewer possibilities among which we may find the one that will open our understanding for the basic complementarity which pervades all of our physical universe.

But we should give Sagredo a chance to comment on this, since his original manner of looking at our problems often gives us a new perspective.

SAGREDO My dear Salviati, you remind me a bit of one of those clever prosecutors who is able to extract almost any confession from an unwitting witness. You have almost convinced even me that the analogy between the random photon arrival sequence and Brownian movement is not a very good one. But should we not go further and admit the possibility at least theoretically, for the sake of argument, that physical reality occurs on different levels and that the uncovering of atomic structures accomplishes only the passage from one level, the classical continuum, to another, the quantal atomic level?

Now on the atomic level we observe unexplained random sequences which seem not to be correlated to anything else that we can observe on this level. Is it then not natural to postulate that there is an entirely new level of reality that we are uncovering here, perhaps very incompletely so far, and that it behooves us to explore other phenomena which might throw light on this possible substratum of reality?

I know that I must sound very vague to you and that perhaps all this may not make any sense at all. I must ask your indulgence because I am not so accustomed as you to the rigors of scientific deductions but am more inclined to imagination and reverie. But Simplicio's concern haunts me, I must admit. I would feel better about the matter if I could see a coherent physical picture which would give us a causal description of the random sequence of photons.

SALVIATI It is hard to disagree with such an eloquent

plea, and there is much common sense in what you say. I am very grateful for your contribution, and I see that our friend Simplicio, too, has recovered from his depression of a minute ago.

SIMPLICIO Sagredo has expressed exactly what I wanted to say in the first place. Only I did it so clumsily that you, Salviati, had an easy time refuting it. But when Sagredo spoke I was reminded of a great teacher and statesman,[9] who taught us the fundamental truth that all events in nature are internally connected, and that it is incorrect to consider things and events as separated from one another. In other words, there are no separate phenomena, but every phenomenon perceived individually belongs to a whole into which it is integrated.

Thus, if we look only superficially we might think that photons behave like capricious individuals, but if we could penetrate more deeply into this inner structure of the world we would uncover their integration into a larger structure of realities, which so far escapes us. Their behavior would then appear to us as the effect of the physical properties of this larger structure.

If we were able to see this connection we would perceive not only the causal connection just mentioned but we would also discover that this causal connection is only a small part of the general network of connections in all parts of the material universe. It would thus infinitely enrich our understanding of all observed and as yet unobserved phenomena.

SALVIATI Tell me, Simplicio, do you believe in astrology?

SIMPLICIO Certainly not! Astrology is a superstition used by the unscrupulous to mold and control public opinion and to exploit individuals. It is unworthy of a scientist to occupy himself with such a fraud. But, pray, why do you ask?

SALVIATI   Because when you spoke I was reminded of a passage from Plato's *Timaeus*.[10] I am not sure whether I remember it correctly, but the argument seems to be essentially the same as yours because Plato considers the universe a single living body in which every part is only a part of the whole. The souls of men are stars which enter man when he is born and return to their places when he dies. Thus man's fate is intimately interwoven with the rest of the universe, in particular with the stars.

SAGREDO   This is true not only for astrology. I have heard that even before the Chaldeans propounded human dependence on the stars many methods of divination were in use for establishing connections between apparently unrelated events. For example, people discovered that when oil was poured on water the various random shapes it assumed in spreading revealed the shape of things to come.

There was also a highly developed technique called hepatoscopy, which was used for the divination of human fate. It consisted of recording and interpreting the seemingly random occurrence of structures, shapes, and lines on the surface of the livers of animals.

Still another form of divination is based on the Chinese I-Ching, the "book of changes," a book that has many followers today. Here the random arrangements of a broken and a connected line representing the two opposing principles of "Yin" and "Yang" are connected with the fates of men.

SALVIATI   These things are far from being understood and to dismiss them as complete nonsense is quite contrary to the scientific spirit. I would add that the Rohrschach test, which in many ways resembles the old experiments with oil poured on water, is a useful diagnostic tool, and is used by almost every psychiatrist today.

So let us take Simplicio's suggestion seriously, and try to work with it, notwithstanding its close affinity with magic and divinatory practices, which may make him less comfortable than he had hoped to be.

Simplicio   I admire your expression of tolerance, but I know from past experience that you feign tolerance in order to destroy all the more effectively a theory that does not agree with your preconceived notions.

Salviati   Now come, Simplicio. Here you are really doing me an injustice! My only motive is to find the truth, in whatever form it may be revealed to us.

Have we not seen hundreds of the most exciting years in the history of science, since our great master, the Academician,[11] who inspired our first conversation, replaced Aristotelian physics by a new one, marking the beginning of the modern era in the history of science?

You remember no doubt how he was fascinated by circles, just as Aristotle, Ptolemaeus, and Archimedes were, and that the very essence of the new physics which he discovered contained the seeds of destruction of this magic spell of the circles.

I sincerely believe that if our great master were present among us, he would be the first to recognize his error. And with his usual originality that startled us so often he would lead us again to deeper, more fruitful ideas.

This is the essence of scientific progress; it can transcend its own boundaries with a process of enlargement of our consciousness that starts with the archetypes of our deepest subconsciousness.

In this region of our souls, rationality is not the main characteristic, as it is on the conscious, scientific level. Here the different functions of our souls are active in undifferentiated form and produce images of symbolic content, which represent essential steps in the process of individualization.

I do not find it dishonorable to recognize the irrational sources of our purest, most fruitful scientific ideas, because life itself and all that it encompasses remain a mystery to us in spite of all the insights we have gained during recent years.

SAGREDO   Your general concept of scientific progress may be beautifully illustrated by the celebrated discovery of irrational numbers by the Pythagoreans. Did they not teach that eternal truths are found in the harmonies of the physical world, expressible in terms of integral numbers and their proportions? When they discovered that the diagonal of the square could not be related in this manner to its side they were deeply disturbed, since this discovery put in question ideological principles which represented the very foundation of their faith.[12]

SALVIATI   Exactly! The discovery of irrational numbers is truly one of the greatest that man has ever made. It alone would grant immortality to the scientist who made it.

So let us not be afraid of ideologies and their systems of images. We should take them seriously, as the symbolic expression of the archetypal structure of the human soul, infinitely rich as a potential source of new scientific concepts.[13]

SAGREDO   This seems to me a fitting moment to close our discussion for today, because your last observation, my dear Salviati, needs to be contemplated by us all. So let us take some time off and relax in the contemplation of the beautiful lake. I see that some light evening breezes have arisen on it, and I invite you to join me on our sailboat.

SECOND DAY

SALVIATI It seems to me that our discussion yesterday may have digressed too far from the path of our discourse, and before we start today's dialogue it might be good to try to recover the essentials of our conclusions.

You, Sagredo, have often done this service for us, why don't you try again?

SAGREDO It will be a pleasure, but I hope you will be patient with me, since I must confess that I left our meeting yesterday in a state of considerable confusion. My mind was disturbed as to the meaning of all those references to the irrational and accidental.

For I have always believed that nature and natural phenomena are essentially intelligible to us, that there are immutable laws which we can uncover by observation and integrate into a coherent system, and that these laws leave no room for ambiguity or hazard.

Yet, we have found that experimental situations exist where the hazard remains in spite of all attempts to control conditions to the maximum possible. You, Salviati, have mentioned the photon polarization experiment, but I have since found by consulting some books that there are a great many other experiments of a similar kind which give similar results.

We have explored the possibility that there are perhaps additional physical properties, represented as "hidden parameters," which might be held responsible for the variation of the outcome in this and similar experiments.

In pursuing this thought further, we concluded that such a description would be possible, but that it would be hard to distinguish from any other kind of magic which relates apparently unconnected random occurrences.

However you, Salviati, to my surprise, suggested that we should pursue this idea more seriously and examine the possibilities of such an approach to the problem.

SALVIATI   There seems to be no question that a suitable "hidden variable theory" of the microsystems is capable of explaining *all* the observed facts. Many people have given examples of such theories, some of them intricate and ingenious, but we need not concern ourselves with the close examination of such examples to convince us of the truth of such possibilities.

It suffices that this possibility is almost a question of definition, since we can simply postulate the completely deterministic behavior of all physical systems, and, whenever a phenomenon appears whose behavior does not conform to this postulate, we introduce a hidden variable for its description. The number of hidden variables ultimately needed may be quite large and may have very peculiar properties, but that does not detract from the possibility in principle. Would you agree with that, Simplicio?

SIMPLICIO   Yes, I agree. In fact we have several examples of such theories.

SALVIATI   Which of these different possibilities is, in your opinion, correct?

SIMPLICIO I have no sufficient reason to prefer one over another.

SALVIATI Do you mean to say that all are equally valid?

SIMPLICIO Every theory I have heard about is capable of reproducing all the known results of quantum mechanics.

SALVIATI Why then should we not continue to work with quantum mechanics?

SIMPLICIO Insofar as the *known* results are concerned, the rules of quantum mechanics would suffice. But it is possible that there are as yet unmeasured effects which would enable us to distinguish between different hidden variable theories and conventional quantum mechanics.

SALVIATI Do you know of any such effects?

SIMPLICIO I personally do not, but for me this question is not of decisive importance, since the recovery of the deterministic behavior of nature makes these theories with hidden variables much more satisfactory to me whatever the ultimate form of the theory might turn out to be.

SAGREDO I have heard that a certain form of hidden variables is actually accessible to experimental tests. They are sometimes called *local hidden variables*, and they lead to predictions which disagree in certain cases with those of ordinary quantum mechanics.

Unfortunately I am not familiar with the details of this theory, nor do I know whether such experiments have been carried out; but you, Salviati, can perhaps explain this to us.

SALVIATI Yes, Sagredo, you are perfectly correct. Experimental tests are possible for some hidden variable theories. Some such tests have been performed and they turned out negative.[1] This means of course only that these particular forms of hidden variables are thereby refuted by their observable consequences; it does not mean that other theories could not be invented which would do the job better.

SIMPLICIO   I quite agree with you, and in fact I believe it will always be possible to find a hidden variable theory that will fit *all* the observed facts and give us a deterministic theory of physics.

SALVIATI   Simplicio, you do not seem to be aware that you have just pronounced the death sentence on hidden variable theories.

SIMPLICIO   How so?

SALVIATI   Don't you know that a theory which can be made to agree with all possible observable facts is no theory at all?

SIMPLICIO   I always thought that this is just the merit of a good theory, to agree with the facts.

SALVIATI   It surely must do that, but it also must have a predictive power for new experiments. A theory which can be adapted to all possible future experiments has no predictive power and is therefore entirely useless.

SAGREDO   When you spoke just now I was reminded of a well-known physicist friend of mine, who works in the field of elementary particle physics. He is very clever and works exceedingly fast, and since theory in this domain is very shaky, for every crucial experiment in process he has four or five theories worked out for whatever possible results the experiments might produce. He keeps these theories hidden and locked up in the drawers of his desk until the experimental result is established. When the result is announced, he draws out the corresponding theory and publishes it quickly. In this manner he is always able to be the first to produce a fitting explanation, and so he has become very famous.

SALVIATI This is indeed a most clever way of doing theoretical physics and its usefulness is increasingly recognized by our younger generation.

SAGREDO I know you too well, Salviati, not to realize that you are being ironic and that you deplore this proliferation of theoretical work as much as I do. If brought to its logical conclusion it will lead to a nightmare of general pollution by a random output of theoretical trash.

SALVIATI You have interpreted my words correctly, Sagredo. The proliferation that you speak of is in my opinion the beginning of the end of our scientific civilization.

Indeed, there is no reason to believe that science, which has had some glorious periods of evolution, separated by vast periods of stagnation, should continue to evolve at the same rate that it has since we met last in 1638.

You remember how optimistic we were at that time, in spite of the formidable handicap of the authority of the Church, which tried to restrict our freedom. Then, it was this very opposition which gave us our invincible force, and the victory we won gave us courage to continue.

Today, with the freedom to do research unchallenged, nothing restricts our scientific production. Yet its quality goes steadily down. With no standards to look up to any more, we are gradually sinking into a morass of scientific gobbledygook.

And yet we have arrived at one of the great moments in physical science because we have uncovered the immensely rich world of microphysics, which puts in doubt almost every concept that we considered firmly established.

SAGREDO I gather from your remarks that you would consider the occurrence of randomness in microphysics as indicating a limit to the applicability of such concepts as determinism.

SALVIATI   That is indeed my feeling, and not only as concerns determinism. The very notion of a system's having a property independent of its relation to the outside world seems to me untenable when one thinks through the vast experimental material on the behavior of microsystems.

SAGREDO   I have an uneasy feeling when you speak like that. We live in a world where real objects surround us with objectively given properties.

In fact, all the information we have on microsystems such as photons, atoms, or elementary particles is obtained by procedures and results in this world of classical physics.

Yet, we maintain that all our macroscopic bodies of classical physics are composed of atoms and elementary particles held together by forces of various kinds. There must therefore exist a boundary where the classical description ceases to have validity and the quantum properties become dominant. Now nobody knows the exact position of this boundary. Most people would agree that the experimental apparatus with which we execute the experiments and the computers with which we evaluate the data are on the classical side, and therefore behave according to the laws of classical physics. But between this input and output there is a system, like the photons in your experiment, which behaves quite differently from any classical system that we know. Thus, by setting a boundary somewhere, on one side of which things are classical and on the other side quantal, we cause almost insoluble problems of fundamental importance.

First, we introduce into the description of nature a dichotomy which destroys the unity of science.

Second, we have no precise notion where this boundary should be placed.

Third, even if we adopt some practical choice for a position of the boundary, we must admit that we have only approximate validity for the physical laws that we uncover with our experimental equipment. Perhaps they are very good approximations, but approximations nevertheless. The question always remains then, what would the exact physical laws look like; could they still be formulated with the notions we have abstracted from the behavior of things around us, which we see, hear, and touch?

SALVIATI I share your uneasiness, and I have often asked myself the same question: Where is the boundary between classical and quantal physics? Until now I have not found a completely satisfactory answer. I suspect the problem is a pseudoproblem and there should be no boundary.

SIMPLICIO I am convinced this cutting of the world into two slices, a classical and a quantal one, is nonsense, and I agree with you wholeheartedly that there should be no boundary.

SALVIATI Our agreement, as usual, is only apparent, my dear Simplicio, since you want to remove the boundary by shifting it to the right in this picture, so that the whole world becomes classical. This you can do only by introducing hidden variables into the description of microevents, a procedure which we have found questionable because it has no predictive power.

I, on the other hand, am inclined to the other alternative,

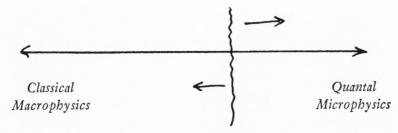

Classical   *Quantal*
Macrophysics   *Microphysics*

BOUNDARY

FIG. 3. *The boundary between classical and quantal physics.*

viz., to shift the boundary to the left, so that all physics becomes essentially quantal, at least in principle.

SIMPLICIO   This is a desperate proposal indeed. It seems like questioning a glorious tradition of classical physics for the sake of saving your particular interpretation of the events that we observe in microphysics. It is contrary to all common sense and the daily experiences of our senses!

SALVIATI   I hope, my dear Simplicio, that you do not mind if I remind you that you used almost identical words when we discussed the heliocentric system about 340 years ago. History has shown that you were in error in a much more subtle way than any of us realized at that time.

So, rather than jump to conclusions, let us examine the difficulties which my proposal would have to face to find out whether they are in fact insurmountable.

SIMPLICIO   I believe they are insurmountable!

SALVIATI   It would be kind of you, Simplicio, if you let us in on the secret which gives you such assurance as to the ultimate form of physical reality.

SIMPLICIO   You always put my convictions to a test as if you yourself did not know the reasons for them. But I am very well aware of your dialectical traps. This time, however, you are up against a formidable opponent. Didn't one of the most famous physicists of the twentieth century proclaim that physics is the science of physical reality and not of our thoughts, much less our sense impressions?[2]

The notion of physical reality is a "category" or a schema of thought which we need in order to render intelligible the totality of the content of consciousness. We cannot give a definition of "reality" without getting involved in circularity, but there is no obstacle to using such a concept as a heuristic principle, justified entirely by its usefulness.

Thus, guided by this thinking, we attribute to physical

objects at every instant a definite position and a definite velocity or whatever physical attributes may be needed for a complete description, even though these quantities may be unknown to us. These quantities evolve continuously in the course of time according to a determined physical law, and our noblest task is to find this law.

This is a challenging program worthy of the greatest minds, and past experience has shown us that along this line true progress is possible.

SALVIATI   I thank you, Simplicio, for your clear and spirited defense of the realistic position with its corollary classical physics.

Would you mind answering some very elementary questions?

SIMPLICIO   Please go right ahead; I shall try to answer them as well as I can.

SALVIATI   Do you know whether the sun will rise tomorrow? [3]

SIMPLICIO   Of course it will.

SALVIATI   Does this mean that it is unthinkable that it will not happen?

SIMPLICIO   No, I can't say it is unthinkable, but it would be most unlikely.

SALVIATI   In other words, the probability that it will rise is so close to one that for all practical purposes it is useless to consider the difference. Would you agree to that?

SIMPLICIO   Yes, I would.

SALVIATI   Would you also agree that it is not absolutely certain?

SIMPLICIO   I think that must be granted too.

SALVIATI   And why would you concede a slight deficiency in absolute certainty?

SIMPLICIO   Not all events in the solar system can be

predicted with certainty since we do not know all the initial conditions with sufficient accuracy to ensure such predictability.

SALVIATI   But if we could know with certainty the relevant initial conditions, then we could also predict the sunrise tomorrow with certainty?

SIMPLICIO   Yes, I believe this to be the case.

SALVIATI   Suppose that at midnight the universal law of gravitational attraction changed its sign, so that it became a law of repulsion. What would happen to the solar system?

SIMPLICIO   It would explode.

SALVIATI   And the sunrise?

SIMPLICIO   Look, under those conditions nobody with a working mind at his disposal would give a hoot about the sunrise. We would all be flying around in space at ever-increasing speeds. I do not see what you want to prove with such a fantastic question.

SALVIATI   Now don't get angry. Your answer proves nothing of what actually would happen; it only proves that such a change in a fundamental law is not unthinkable and it seems to me therefore logically possible.

The reason you think it fantastic is that we rarely observe such changes. Or, to be more precise, if there were such changes, nobody would be here any more to record them, and therefore we are used to discarding such possibilities as absurd.

But that they are not entirely absurd can be seen from the fact that some people have seriously discussed the question of whether the fundamental constants of physics, such as the speed of light, the electric charge, Planck's constant, or the gravitational constant, might not vary with time.

There is not such a radical difference between a slowly varying function and a discontinuous function.

Once the principle of a variation is admitted, the spell of the sanctity of physical laws is broken, and anything might be possible.

SIMPLICIO   These seem to me idle speculations and not much good can come of them.

SALVIATI   You forget, Simplicio, that we are examining your thesis that physical laws must be deterministic, and we have reached the conclusion that such a statement can be decided ultimately only by an appeal to experience and not by an appeal to reason.

SIMPLICIO   I agree with that.

SALVIATI   Once you agree to that it remains to be seen what experience can tell us about your faith.

SIMPLICIO   I believe the best confirmation of this thesis comes from the fact that it leads to predictions that can be verified.

For example, when the calculations of Leverrier indicated that the irregularities in the movement of Uranus were due to the presence of an unknown planet, Neptune was discovered at the exact place where the calculations predicted it to be.

SALVIATI   There is no doubt that such experiences lend credence to the belief that the laws of evolution of physical systems are completely deterministic, so that the future can be predicted with certainty from the present state, provided this state is sufficiently well known.

What disturbs me a bit is that we cannot formulate this belief of yours very easily except in anthropomorphic terms: *Knowledge* of sufficient properties at one time permits *prediction* of other properties at other times.

What does our *knowledge* have to do with the physical laws anyway? Could we not find a formulation which avoids such a reference to *knowledge* and *prediction*?

SIMPLICIO   I think such a formulation is completely equivalent to another one, free of this objection:

Similar physical situations are followed by similar effects.

SAGREDO   May I intervene here for just a second to remark how closely you two have duplicated the train of thought which led David Hume to his definition of causality. But I suppose you are both aware to what enormously difficult philosophical questions this leads.

SALVIATI   Your warning, Sagredo, is much to the point, so let us beware of philosophy and just examine whether Simplicio's latest formulation can be used to support his view of the deterministic nature of physical laws.

In order to pursue this point I brought along a pinball machine. Both of you have played similar machines, I am sure, so my demonstrations will certainly not surprise you.

Normally the player tries to activate all kinds of springs and levers in order to keep the ball productive as long as possible.

To demonstrate my point I shall omit the player entirely, and just activate the release of the ball with this mechanical device, which is so constructed that it reproduces as nearly as possible the same initial conditions.

So here we go; I start playing and the automatic counter sums up the score. We have reached 17,300 for our first play. We continue the play several more times and keep track of the scores: 13,000, 7,200, 14,000, 2,500, 19,700, 16,500, . . . , and so on.

You notice no doubt that we obtain a different score every time we play. In fact it looks suspiciously as if this difference has a random fluctuation around an average value.

How can this be brought into agreement with your law of causality, Simplicio?

SIMPLICIO   You know, of course, that the initial conditions for each play are not exactly the same, so it is not surprising that the result is not the same either.

SALVIATI   Nothing that happens in the world at any given time is exactly the same at another time. This is granted. But should we not say that the conditions are *nearly* the same?

SIMPLICIO   I think we should.

SALVIATI   And that the effects are vastly different?

SIMPLICIO   Undeniably so.

SALVIATI   So the law of causality, as you announced it just a while ago, does not hold in this case?

SIMPLICIO   I suppose one is forced to this conclusion.

SALVIATI   Would you also agree that here we are in the realm of classical mechanics, and that there can be no question of quantal properties entering into the description?

SIMPLICIO   Yes, of course.

SALVIATI   So we have here evidence that even in the classical realm strict causality or determinism cannot be proved in all circumstances?

SIMPLICIO   I think that must be granted.

SALVIATI   May I remind you that only a few minutes ago you made a declaration of faith, asserting essentially the very thesis that we have now recognized as being neither logically necessary nor empirically established.

SIMPLICIO   I know that, but I must say that you chose the worst possible example. From a mechanical point of view a pinball machine is an exceedingly complicated system, and the equation of motion for the descending ball, although certainly classical and deterministic, cannot even be written down, let alone solved.

However, I believe that if you choose a sufficiently simple system you could demonstrate causal evolution in the state in accordance with my definition of causality, which you, Sagredo, have attributed to David Hume.

SAGREDO   I believe that you are mistaken, Simplicio. Several years ago I heard a lecture by a very prominent physicist[4] who demonstrated to us that even for very simple systems one runs into the same sort of problems as with the pinball machine.

SIMPLICIO   This I would like to see.

SAGREDO   I am not sure that I can reproduce everything exactly as I heard it, but I believe I can give you the main point of the discussion.

Let us consider for our system a simple wheel mounted without friction so that it can turn freely around its axis, which passes through the center of the wheel.

If we set the wheel in movement by giving it an initial turn, it will keep on moving indefinitely.

The state of the wheel at any time is given by its position and its velocity of rotation. The positions can be fixed by choosing an arbitrary direction in the plane of the wheel, and by giving the angle ϕ of a fixed direction on the wheel with respect to the fixed direction in space.

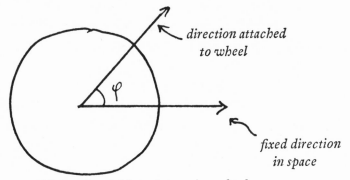

FIG. 4. *Fixing the position of a moving wheel.*

Since by hypothesis the wheel moves without friction, its speed of rotation ω will always remain the same. But its position $\phi(t)$ will constantly change according to a law which we might write as follows:

$$\phi(t) = \omega t + \phi_0.$$

Actually this equation is a bit misleading since the position of a wheel is completely determined if we know the angle ϕ within the range $0 \le \phi < 2\pi$. The angle ϕ should be reduced to this range by subtracting as many multiples of 2π as necessary. The mathematicians would say that we consider the angle $\phi(t)$ only "modulo 2π."

Suppose now that we make a very small change in the initial velocity, that is in the constant ω. Let us say that we change it to a slightly different value $\omega' = \omega + \Delta\omega$. The angle $\phi(t)$ will then change to a slightly different angle

$$\phi'(t) = \omega't + \phi_0,$$

and we may calculate the difference between the new angle and the old one by taking the difference of the last two equations

$$\phi'(t) - \phi(t) = \Delta\omega t.$$

Now no matter how small $\Delta\omega$ may be, we can always choose t sufficiently large (that means we wait sufficiently long) to obtain any value we wish for this difference. Hence for times of the order $T \simeq 2\pi / \Delta\omega$ or greater, the position of the wheel is completely undetermined.

Formulated differently, no matter how small an uncertainty we choose for the initial velocity, if we wait sufficiently long, the position of the wheel will be completely undetermined.

Thus even in this simple example, the strict causality as

you formulated it, Simplicio, does not seem to be applicable.

This makes me wonder whether there is any example at all where it is applicable.

SALVIATI Your example, Sagredo, of an exceedingly simple but dynamically unstable system, is indeed a most valuable contribution to our discussion, and I am deeply grateful to you for bringing it to our attention. It illustrates better than anything I might have said, how essential and fundamental is this point in classical mechanics.

I see that our friend Simplicio is hard pressed to find a way out of this difficulty. Do I interpret your feelings correctly? Do you have any reply to this, or are you ready to admit that your faith in the deterministic character of classical evolutions cannot be established on empirical grounds?

SIMPLICIO I agree that your examples show that the belief in determinism cannot be established with certainty, but I still maintain that there is evidence that such a belief makes good sense as a limiting case.

In the past have you not often referred to the words of the Academician[5] who inspired our previous discussion, admonishing us that we should be able to see behind the phenomena and not cling to inessential effects which have nothing to do with the question under discussion?[6]

I was thinking especially of the case of the free fall of heavy objects, where we Aristotelians objected to his theory since he always wanted to exclude the resistance of the air, while we considered it essential to the problem. I have since made much progress; I have learned his lesson well, and I do think it is applicable again in the point under discussion.

SALVIATI How stubborn you are, Simplicio! I cannot help admiring you! Your objections are a challenge to us to think more deeply about the foundations of physics, and so I think we should all be grateful to you.

So let me take up Sagredo's beautiful example of the wheel, and let me ask you some questions about it.

Suppose that the starting angular velocity is uncertain within a range of $\Delta\omega$. How long would you have to wait for this uncertainty to make itself felt in the position of the wheel?

SIMPLICIO   It seems to me we would have to observe as long as necessary to make the uncertainty in the position easily measurable, for instance of the order 2π.

SALVIATI   How long would that take?

SIMPLICIO   According to Sagredo's formula we would expect this to be of the order $2\pi/\Delta\omega$.

SALVIATI   And for a time much *shorter* than that you could say nothing about the uncertainty?

SIMPLICIO   Nothing.

SALVIATI   Then in that case the uncertainty in the initial velocity could be much larger than $\Delta\omega$ and you could not detect it through an observation of position.

SIMPLICIO   Yes.

SALVIATI   Could we not say this also in the following way? In order to determine an uncertainty in the angular velocity of order $\Delta\omega$ we need to observe the wheel a length of time of about Δt, where

$$\Delta t \, \Delta\omega \gtrsim 2\pi.$$

SIMPLICIO   This is a concise way of saying the same thing.

SALVIATI   Does it not follow then that in order to determine the initial angular velocity with an arbitrary degree of precision, we would have to make a measurement which lasts an infinite amount of time?

SIMPLICIO   This is surely an inevitable conclusion.

SALVIATI   Is it not evident that such measurements are in principle impossible?

SIMPLICIO   I think that is true.

SALVIATI   So that any discussion based on the hypothesis of the existence of such data is really just empty talk?

SIMPLICIO   I do not know what to say to this.

SAGREDO   Salviati, I must admire you. For what you have done with my simple example is quite astonishing, and I dare say brings to mind all kinds of ideas, not the least of which is the feeling of how much more satisfying it is to the mind to contemplate circular movement than movement in a line.[7]

In circular movement we seem to have brought eternity into the confines of our limited comprehension. It is both eternal and confined, hence offers a glimpse of an eternal truth which the Newtonian tradition has neglected.

Perhaps the fascination which circles held for our Academician showed a deeper insight than his interpreters were willing to grant.

But let me mention another idea: Your formula for the uncertainty of angular frequency and time has a close similarity to another formula often mentioned in wave mechanics. If we use de Broglie's relation between energy and the frequency for a circular movement we obtain for the uncertainty of the energy

$$\Delta E = h \Delta \omega / 2\pi$$

so that from your inequality, Sagredo, I obtain

$$\Delta t \, \Delta E \gtrsim h$$

(where h = Planck's constant).

This is one of Heisenberg's inequalities, which has been

much discussed in connection with the foundation of quantum mechanics.

I do not know for sure how to interpret this inequality, but I am fascinated by the close affinity of the two inequalities. Yours, Sagredo, seems to be valid for the classical domain, and the passage to the quantum domain requires nothing more than de Broglie's relation.

SALVIATI   Indeed, Sagredo, I am pleased because one of my motives in discussing this example with Simplicio was to bring out, more than is usually done, the close affinity of the epistemological situation in the quantum and the classical realms.

Let me illustrate this by another example: Once we have granted the fundamental impossibility of attributing any precise values of the mechanical quantities to any classical system we are better prepared to accept another description of the state of a physical system which seems to be in better agreement with this fundamental fact.

Instead of giving the state of position and motion of our wheel at the time $t = 0$ by an exact position $\phi_0 = \phi(0)$ and an angular velocity $\omega_0 = \omega$, we may give a more general description in the form of probability distributions, let us say $\rho_1(\phi)$ and $\rho_2(\vartheta)$, where $\rho_1(\phi)d\phi =$ the probability that the angle has a value ϕ within $d\phi$, and $\rho_2(\omega)\,d\phi =$ the probability that the angular velocity has a value ω within $d\phi$.

It follows from this definition that ρ_1 and ρ_2 are two positive functions which satisfy the normalization conditions

$$\int_0^{2\pi} \rho_1(\phi)\,d\phi = \int_{-\infty}^{+\infty} \rho_2(\omega)\,d\omega = 1.$$

We could of course do this a bit more generally by introducing the notion of probability measure, so that the limiting case of the exact value of position and velocity could be handled in the same manner. But complete generality is not the point now.

Such density distributions ρ_1 and ρ_2 are then the mathematical representation of systems (in this case the wheel) that we have prepared in some way or other. They represent the best we can say about the initial conditions that we can reach with any physically possible preparation.

On the other hand, if we are given these quantities and the equation of motion $\phi(t) = \omega_0 t + \phi_0$, we can calculate these probability distributions at any later time. This seems to me a model for the formulation of mechanics which is much closer to the actual physical situation.

SIMPLICIO   I feel that this formulation raises a difficulty which you have avoided mentioning.

SALVIATI   And what could that be?

SIMPLICIO   You speak of probability when you define the functions ρ_1 and ρ_2. But the probability of an individual object does not have a well-defined meaning. We can speak of probability only for a member of an ensemble. Thus with your last formulation of mechanics you no longer have a theory for an individual system but only for an ensemble of suitably prepared systems.

SAGREDO   I think Simplicio is right. In fact when he spoke I was reminded of some of the well-known problems connected with the interpretation of probability statements.

An individual can be a member of many different ensembles and his probability distributions are largely undetermined unless the ensemble of which he is a member is specified.

SALVIATI   I have no objection to this statement. Indeed I never believed that the dynamical theory can be considered strictly a theory for an individual system.

SAGREDO   But this is certainly not what is taught in the books on classical physics.

SALVIATI   Yes and no. You are right insofar as the language people use is concerned, but I do not think there is disagreement on the essence of the matter.

Almost everybody would agree, I believe, that the laws of classical mechanics do not describe the evolutions of the state of an individual system. They give a general schema of evolutions which is applicable to a general class of individuals. In order to obtain the actual evolution of an individual one must specify the initial conditions.

SAGREDO   Now I understand why Newton needed a God to put his planets in motion. It is evident that his laws of motion could give him many other solutions for the planetary orbits which are not observed. There is nothing in these laws which would require these particular solutions with the exact number and size of planets as they are observed. And is it not natural to assume that the particular arrangement of the planets with nearly circular orbits, all in the same plane and with the same direction of motion, shows the hand of the original designer?

SALVIATI   Yes, that was Newton's quest. And I believe he was deeply aware that every real event is something single and unique, which in a strict sense never happens again. Some of his followers may have forgotten that important point. It is the very essence of a physical law that it refers to reproducible physical situations and therefore, strictly speaking, never to an individual event, or rather to individual events only insofar as they are members of an ensemble of similar events.

SAGREDO   Since we are tired I do not want to prolong this meeting any longer, except to say that now I understand for the first time why the uniqueness of the individual is incompatible with scientific statements that you might want to make about such an individual.[8]

Do we not see here another example of that all-pervading principle of complementarity which excludes the simultaneous applicability of concepts to the real objects of our world?

Is it not so that, rather than being frustrated by this limitation of our conceptual grasp of the reality, we see in this unification of opposites the deepest and most satisfactory result of the dialectical process in our struggle for understanding?

But, my friends, I do not want to keep you longer. Be of good cheer and enjoy the sights and pleasures of our lovely city until we meet again tomorrow at the appointed hour.

THIRD DAY

SAGREDO Forgive me, Simplicio, I do not want to intrude with personal questions, but you worry me today. You don't look well. Are you feeling the after effects of night life in Geneva?

SIMPLICIO It is true I feel somewhat strange, but it is not due to the excesses of pleasure in your delightful city. On the contrary, I returned before the night was well advanced in order to meditate more about our discussion yesterday, but before my thoughts had reached any definite focal point, I must have fallen asleep and then I had a strange dream.

When I awoke I was tense and confused and I could not find peace of mind. So I lay awake for the rest of the night. This explains my fatigue.

I felt somehow that the dream must have something to do with the subject under discussion, but it was so strange that no matter how I tried I failed to make sense out of it.

SAGREDO Could you tell us the dream so that we may think about its significance with you?

SIMPLICIO I do not mind telling you my dream since you are interested, but I doubt very much that you can extract any profit from it.

· *49* ·

SIMPLICIO'S DREAM   I was in a vast, dimly lit hall, like a church, and there were very many people around. Someone was officiating as at a religious service,[1] but I could not see him at first because the crowd barred the view. The people were all very solemn and they seemed to know what was going on,[2] but I was completely in the dark as to the meaning of it all. Finally I asked somebody next to me what this was all about. He looked at me rather surprised and then he said in a low voice, "Don't you know that we are here to await our fate?"[3]

This did not make much sense to me, but because of the solemnity of the occasion I did not have the courage to ask further.

Then I noticed that the crowd was slowly moving toward the center of the sanctuary. I moved along with it, and when I finally arrived close enough to see, I perceived a very old man[4] with a silvery beard wearing a long white dress. He was standing near a roulette table throwing a ball made of luminous stone.[5] Many players stood around the table. Each one had only one chance to play. After the play a little Maxwell demon[6] handed each player an envelope containing his fate and escorted him from the room.

When it was my turn I was frightened and wanted to run away. But one of the Maxwell demons had put his crooked hand around my wrist and croaked with a fantastic little giggle: "Come on, buddy, you can't escape your fate."

Seeing that there was nothing else to do, I took my chip and put it on number three. A woman who was standing next to me whispered in my ear[7]: "You are a fool, you will never unify the opposites on three. You should have chosen four."

I felt terribly foolish at not having noticed that myself[8] but at this very moment the old man said sharply: "Les jeux sont faits, rien ne va plus."

The luminous ball was circling around in an erratic manner, tumbling from one number to another, until it finally came to rest on number three. I had won![9]

A muffled exclamation of wonder broke the silence and everybody looked at me with astonishment and envy. The mean little Maxwell demon became excessively polite, and the croupier handed me a golden[10] envelope, which contained a card with the following inscription:

The Green Man will see you in his chambers; follow your guide.[11]

Immediately my little demon took me gently by the hand and led me through the crowd, which cleared the way respectfully to let us pass. We came to a door and then decended a spiral staircase into a long, dark passage. At its end we entered a chamber lighted with a pleasant, pale green light. At the wall opposite the entrance was a throne, on which sat the "Green Man," who immediately addressed me with these words: "I have expected you for a long time and you never came, but now that you are here, it is time to begin. *You must find the unknown road to truth.*"

I was deeply impressed and did not know what to say. Before I could collect my thoughts the "Green Man" continued: "Tell us what you need in your search for truth, it shall be granted. But remember, you can have only one wish, the rest is up to you."

I was so startled by this sudden revelation that I could scarcely think. Then the meaning of the statement became clear to me, and I said quickly: "Give me a library containing the books with all the wisdom, all the truth, and all the beauty of all time."[12]

I had barely finished speaking when the light in the room

dimmed and a strange transformation took place. When I began to discern the details of my surroundings I discovered that I was in a kind of open spaceship floating through an immense row of bookstacks, which extended to infinity in all directions. The spaceship could be moved at any speed up to that of light and could be stopped in an instant without the least feeling of discomfort. Every hundred miles in each direction there was a small platform on which was installed a librarian working at a desk. I went to one of the stacks and opened a beautiful, leather-bound volume. To my astonishment I could not read it, since it seemed to contain only a random series of letters and spaces. Thinking that it might be written in a foreign language I went to the nearest librarian, about 30 miles away, and asked in what language it was. His answer startled me even more. He said:

"These books contain everything that ever was and ever will be written in any language of the past or the future. But together with all the meaningful statements, they also contain all the random sequences of letters up to the longest book that will ever be written. This is the only truly complete library in the world because it contains everything."[13]

"But such a library must be infinitely large!" I exclaimed.

"No," he replied, "this library has a finite number of volumes. The exact number is unknown but it is finite."

"But how can I ever find anything in this library?" I asked in despair.

"We have a most efficient electronic retrieval system," he said, "and anything you want can be instantly commanded by composing on this command console a message to the central memory device." To show me how it worked, he commanded for me in an instant a letter written by Galileo to his daughter, Sister Maria Celeste, in 1633.

I was deeply impressed, because I knew that this letter, hitherto lost,[14] contains the clue to the unsolved mystery of Galileo's trial before the Inquisition in 1633.

But before I had time to open the book and read the letter the librarian continued, "Anything that you wish to see is here at your disposal. Try it for yourself."

I thought quickly what I could command, and after some reflection said, "Let me see the theory of elementary particles which explains all known facts about them."

"Which one?"

I was a bit surprised at this question and replied, "I did not know there are several. Of course I want the correct one, that is the one which agrees with all the facts known today."

He smiled and explained, "There are 137 different theories available which satisfy this requirement. You must give me further specifications if I am to select one. Or do you wish to see them all?"[15]

Much surprised that there should be so many different, correct theories, unable to think of any other criterion for selecting one from all this wealth of theories, and lacking the inclination to study them all, I replied, "No, not just now, I just wanted to know what is available."

As he turned away to resume his work, which no doubt was immense, he said courteously, but a bit dryly, "Any time you need anything, I am at your disposal."

I left him, and a deep feeling of depression came over me as I moved aimlessly through my immense library in the three dimensions of never-ending space. . . . . .

SAGREDO   I can well understand your feeling, Simplicio, and the vivid recital of your most significant dream has affected me deeply. As is so often the case in matters which transcend our understanding, we must, I believe, let it rest for the time being, while gradually assimilating the images of your dream into our consciousness.

I therefore suggest that we return to our topic of discussion.

We reached the conclusions yesterday that chance is everywhere in nature and that we have no evidence in any branch of science that things happen with certainty. Nevertheless we agree that some things happen with such high probability that for all practical purposes it is reasonable to suppose that they happen with certainty.

It seems then that the acid test of any science is to be able to make statements about the occurrence of events whose rightness is overwhelmingly probable.

SALVIATI   I agree with you, Sagredo, and I am inclined to believe that once this point of view is admitted, the difference between classical and quantum physics becomes much attenuated. What seemed before an almost unreconcilable division of the world into two opposite camps now becomes more like complementary aspects of one and the same object.

SIMPLICIO   Your way of expressing this is a veiled attempt to reintroduce into our science the thoroughly discredited positivistic philosophy![16] I maintain, together with some of the greatest scientists today and of all time that science is concerned with the properties of *real* objects. The fact that our science is not in a position to make statements about these objects with absolute certainty, does not permit you to conclude that these objects do not *have* definite properties. It simply means that our methods are not sufficiently precise to reveal these properties. This confusion between what things *are in themselves* and what *we can know* of them is quite an elementary confusion.

SAGREDO   I see that in spite of his dream, Simplicio is still in the best of form and his reasoning seems sound. His point of view certainly agrees with all our common-sense experiences.

I had a friend who was struck by an automobile and severely injured. While in the hospital recovering from his accident he reviewed all the city ordinances concerning traffic regulations at the point of his accident. He discovered that no automobile had any right to travel at that point at the time of his accident. He therefore concluded that the accident could not really have happened, and that it was all the result of some collusion of hazards, a random fluctuation perhaps, or a figment of his imagination.[17]

But let me not dwell too long on such an illustration of the obvious, because I for one am most anxious to hear what Salviati has to say to Simplicio's statement.

SALVIATI   I am afraid that I, like you and everybody else, should laugh at the poor fool who thought the automobile accident could not have happened because according to the laws that we have made it *ought* not to have happened. We cannot legislate the laws of nature. Things happen in part independent of us, or rather I should say sometimes in spite of us.

But this is not the question at issue. What Simplicio is talking about can be condensed into one word: the *Real*. He uses this word as if everyone knew exactly what it means, while I must admit that whenever I try to understand its meaning it escapes me at the last moment.

I know that many philosophers use this concept very freely but I have had bad experiences with them. When one asks them what they mean by it, they treat one like a half-wit, just about ready for special care; or they give explanations in such a complicated, technical jargon that no one is the wiser. If one persists with questions they become aggressive because they are very touchy on this subject.

So rather than consult the philosophers let us try to see by examining some special cases, what Simplicio means. You, Simplicio, are questioning the statistical character of probabilistic laws of physics because you cannot conceive that such laws should be the ultimate laws for an *individual* system. That is why you must attribute to an individual system definitive properties, which, although unobservable in practice or in principle, will guarantee for you a deterministic evolution of the state.

But you seem to overlook that this point of view cannot be carried through consistently. I am sure that you know this and that you have merely forgotten it. Let me help you remember:

You both remember our discussions in 1630, when we were comparing the two major systems of the world. Do you think, Simplicio, that our discussions decided the question one way or another?

SIMPLICIO   No, I don't think that we could at that time really decide the question, since we did not have enough data to confirm one or the other of the two systems. All we could do at that time was to show the fallacies of some of the reasons which were adduced in favor of the Ptolemaic system, for we did not have really sufficient reasons to prove the Copernican system. I, at least, was never convinced of your theory of the tides, which you constructed so cleverly in order to demonstrate the daily rotations of the earth.[18]

So at that time we could only say that the question was undecided and since there was this doubt, Cardinal Bellarmine was quite justified in saying that in this case one should not abandon the interpretation of the Holy Bible, transmitted to us by the fathers of the Church.

SALVIATI   But how is it now with you, Simplicio? Have you now decided which of the two views is correct? Which is at rest, the sun or the earth?

SIMPLICIO   I think everybody knows now that Copernicus was right. We even teach it to our children in school.

SALVIATI   I did not ask you what everybody else believes to be true, and what we teach to our children. I asked you what you, Simplicio, believe to be true.

SIMPLICIO   I believe that the heliocentric system is true.

SALVIATI   But you know of course that the sun is not at rest in the center of the universe since it moves inside the galactic system.

SIMPLICIO   Yes, of course I know that.

SALVIATI   And that the galactic system is in motion too with respect to extragalactic systems.

SIMPLICIO   This, too, I know.

SALVIATI   So your knowledge today is quite different from what Copernicus believed and what our Academician tried to prove as real with his ingenious theory of the tides.

SIMPLICIO   It certainly is different. But that is not the meaning of the Copernican theory as it is understood today.

SAGREDO   What is this meaning then?

SIMPLICIO   That the earth cannot be at rest in the center of the universe. And this I believe to be true.

SAGREDO   When you make such a firm statement, you must have some strong reasons for believing it. Tell me, if I were to question this statement today, what could you give as reasons why the earth cannot be at rest?

SIMPLICIO   You are really peculiar! Three hundred

and forty years ago you argued with so much power on the other side that our Academician had the hardest time proving to the Holy Office that he did not believe it! And now you ask me to take up your role of the past!

SALVIATI   I do not ask such a difficult thing of you, although I know that you have learned a lot since we last discussed the matter. I think my question was really unfair since I asked you the impossible.

SIMPLICIO   You really do confuse me now. If that was your intention, you have succeeded. It is up to you to explain yourself.

SALVIATI   The explanation is simple. We owe it to one of the greatest physicists of the twentieth century.[19]

He discovered that, because of the remarkable fact that the gravitational and the inertial mass are exactly proportional in all material bodies, it is possible to develop a general theory of relativity, which includes all kinds of accelerated motions by combining them in an appropriate manner with the forces of gravitation.

According to this theory it is then possible to consider the earth at rest as an equally valid and therefore equally real description of the universe as any other, for example, the Copernican one.

SIMPLICIO   I know this theory, but somehow I have never believed that it throws any doubt on the reality of the motion of the earth around its own axis and around the sun.

Certainly you do not question that it is much simpler to use the Copernican description.

SALVIATI   I certainly agree, but, Simplicio, are you prepared to accept a vague property like *simplicity* as a sufficient criterion for the *reality* of something?

SIMPLICIO   In the case of cosmology I certainly cannot think of another criterion.

SAGREDO We have come a long way since 1630. I certainly never thought then that our friend Simplicio would ever make such a confession. But surely we need not stop here. Cosmology does not seem to be especially suited for coming to grips with reality. Let us take something more down-to-earth.

I suggest that we consider the electromagnetic field. Faraday suspected it, Maxwell discovered its laws, Hertz measured it, and a large industry is based on it. We produce it in power stations, transport it, and sell it for good cash. Anything that can be bought and sold seems to me real; sometimes I almost suspect that these are the only real things in the world. Salviati, what do you say to this?

SALVIATI Indeed it is an excellent example. Let me ask you, Simplicio, do you think a theory of electromagnetic interaction is possible without supposing the reality of the electromagnetic field?

SIMPLICIO Such a theory could certainly not be correct.

SALVIATI And if I tell you that such a theory exists and that it is in perfect agreement with all the known facts of the classical theory of the electromagnetic field, would you still be of the same opinion?

SIMPLICIO I have never heard of such a theory.

SALVIATI One does exist. It is called the theory of *action at a distance*. It has only particles as *real* entities and it uses no field. This theory is in complete agreement with the other theory which you know and which we all use, but it is a bit more complicated. Hence it is less familiar. However, its fundamental ideas are as easy as they are significant.

SAGREDO Please tell us more about it.

SALVIATI With pleasure, the more so as it throws some significant light on the question of *Reality* in physics.

The idea originated in 1845 with Gauss, who introduced the concept of an action at a distance propagating with finite velocity as a fruitful generalization of Newton's theory. It was put into mathematical formulation by Schwartzschild, Fokker, and Tetrode about fifty years ago.

The basic idea in this view is that a radiating system such as an antenna never emits energy except to a receiving system. In this theory it would thus be absurd to think of light emitted by one atom regardless of the existence of an absorbing atom. Instead of *emission* and *absorption* of radiative energy one has only *transmission* of such energy. The conceptual basis of this system is therefore simpler since it contains fewer elements. Therefore, according to your criterion of reality, Simplicio, it should be more real.

The remarkable fact is that such a system can be developed in full agreement with the classical theory of the electromagnetic field.

SAGREDO   I am most impressed by your statement, Salviati. I do not know this theory at all, so you will forgive me if I remain skeptical. But I am sure that you studied it and that you are telling us the truth.

But please tell me, once one has discovered a theory which contains only particles, would it not be natural to imagine that there should also exist a theory which contains only fields?

SALVIATI   This is quite correct. Such a theory exists, at least for the gravitational field. I am sure it is also possible for the electromagnetic field, but as far as I know it has never been fully developed.

Some day somebody may even develop a theory which refers only to *geometry* as the truly *real* element of nature and that regards all other "observed" physical objects, such as particles, fields, energy, etc., as merely different manifestations of one and the same underlying physical reality, *viz.*, geometry.

Would it not be a most satisfying economy of concepts if fields and particles were not foreign objects moving about in the arena of space and time but rather were in some way constructed out of pure space? As one of my friends, who had thought about this matter more deeply than anyone, once put it:

"Is the metric continuum a magic medium which, bent up in one way here represents a gravitational field, rippled in another way there describes an electromagnetic field, and twisted up locally describes a long-lived concentration of mass energy? In other words, is physics basically a matter of pure geometry? Is geometry only an arena or is it everything?"[20]

Personally I do not believe that one necessarily gains a deeper understanding of physical phenomena from such mental acrobatics, but for us the fact that this possibility exists is of fundamental importance because of the light it throws on the notion of *Reality* in physics.

SAGREDO You have given us much to ponder. For one thing, it occurs to me that if reality is already such an elusive concept on the level of macroscopic phenomena, how much more questionable will it appear when we examine phenomena which originate with microphysics?

It seems to me now that reality in itself can have no definitive meaning unless it is tested within a framework of theoretical constructs. It is just like motion in absolute space. Motion acquires its reality only when it is considered with respect to some other object.

SALVIATI This I believe to be the case, and I would

add that these theoretical constructs are, in a much larger measure than is usually believed, free inventions of the human mind. They are not imposed on us by necessity from outside; rather they are derived from images with a long and largely unconscious history. They are what Jung referred to as archetypes of our souls, and therefore belong to a different level of "reality."

SIMPLICIO This excursion into psychology is largely irrelevant for understanding the meaning of the *real* in physics. You, Salviati, have made much of the fact that our theoretical constructs are not *real* in the sense that I understand the term, but this is misleading. We all know that we have no direct access to reality, except by the intermediary of our senses. Furthermore theoretical constructs like energy, fields, particles, etc., are certainly not the reality we seek to understand. These constructs and their mathematical expressions are symbols for some kind of reality which lies beyond the level of everyday experiences.

SAGREDO In other words, the real objects are just like the constituents in a democracy, while the symbols are their deputies in the government.

SIMPLICIO That is an excellent illustration of what I am trying to say. Thank you very much for illustrating it so concisely.

And just as the deputy may change after an election, so these symbols and constructs may change. But the constituents which they represent are the same, and similarly the underlying reality must be the same.

SALVIATI I could go along with this except that you may be referring to a wholly imaginary constituency since you have no way of ever perceiving it in any manner whatsoever.

But I suggest that we do not dwell longer on this point,

for here begins the hunting ground of the philosophers. I am reluctant to trespass on their territory. For us there is still a formidable problem to be solved.

SAGREDO   What is that?

SALVIATI   It has to do with Simplicio's dream.

SIMPLICIO   What do you mean?

SALVIATI   When you told us your dream, Simplicio, I was impressed by the fact that it referred to a point which is fundamental in scientific epistemology, namely the significance of *abstraction*.

SAGREDO   Could you explain yourself a little more clearly?

SALVIATI   Suppose I give you two sequences of numbers, such as

785398163397448309615660 84 ...

and

$$1, -1/3, +1/5, -1/7, +1/9, -1/11, +1/13, -1/15, \ldots$$

If I asked you, Simplicio, what the next number of the first sequence is, what would you say?

SIMPLICIO   I could not tell you. I think it is a random sequence and that there is no law in it.

SALVIATI   And for the second sequence?

SIMPLICIO   That would be easy. It must be $+1/17$.

SALVIATI   Right. But what would you say if I told you that the first sequence is also constructed by a law and this law is in fact identical with the one you have just discovered for the second sequence?

SIMPLICIO   This does not seem probable to me.

SALVIATI   But it is indeed so, since the first sequence is simply the beginning of the decimal fraction of the sum of the second. Its value is $\pi/4$.

SIMPLICIO   You are full of such mathematical tricks,

but I do not see what this has to do with abstraction and reality.

SALVIATI   The relationship with abstraction is easy to see. The first sequence looks random unless one has developed through a process of abstraction a kind of filter which sees a simple structure behind the apparent randomness.

It is exactly in this manner that laws of nature are discovered. Nature presents us with a host of phenomena which appear mostly as chaotic randomness until we select some significant events, and abstract from their particular, irrelevant circumstances so that they become idealized. Only then can they exhibit their true structure in full splendor.

SAGREDO   This is a marvelous idea! It suggests that when we try to understand nature, we should look at the phenomena as if they were *messages* to be understood. Except that each message appears to be random until we establish a code to read it. This code takes the form of an abstraction, that is, we choose to ignore certain things as irrelevant and we thus partially select the content of the message by a free choice. These irrelevant signals form the "background noise," which will limit the accuracy of our message.

But since the code is not absolute there may be several messages in the same raw material of the data, so changing the code will result in a message of equally deep significance in something that was merely noise before, and *conversely*: In a new code a former message may be devoid of meaning.

Thus a code presupposes a free choice among different, complementary aspects, each of which has equal claim to *reality*, if I may use this dubious word.

Some of these aspects may be completely unknown to

us now but they may reveal themselves to an observer with a different system of abstractions.

But tell me, Salviati, how can we then still claim that we *discover* something out there in the objective real world? Does this not mean that we are merely creating things according to our own images and that reality is only within ourselves?

SALVIATI   I don't think that this is necessarily so, but it is a question which requires deeper reflection. We might take it up in a subsequent meeting.

Since I see that Simplicio is tired, I propose that we close our discussion for today.

SAGREDO   I shall be most eager to find out more about the reality of the world, especially as it concerns micro-systems.

FOURTH DAY

SALVIATI Greetings, my friends, and welcome to our final debate on the question, "Are Quanta Real?"

I do hope that you have not become too tired of this subject and that you will not desert me now that we are on the threshold of a deeper understanding of the meaning of reality in physics.

With your permission, I shall not dwell on preliminaries but will begin with the heart of the subject. To this end I brought with me a photograph of a bubble chamber picture, which I borrowed from one of my friends at CERN.

The picture shows the creation of a neutral Λ^0 particle at point 1 by the incident particle from the left and its subsequent decay into two particles of opposite charge at point 2 a few centimeters to the right. The dotted line joining 1

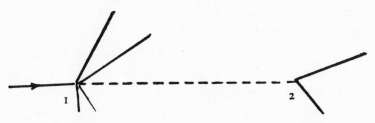

FIG. 5. *Bubble chamber picture of the decay of a neutral* Λ^0 *into* p *and* π^-.

with 2 is not visible on the photograph; it indicates the presumed trajectory of the neutral particle between its point of creation at 1 and the point of its decay at 2.

Such events have been observed many thousands of times and give an accurate determination of the average lifetime of a Λ^0 particle for this particular decay mode.

Now, Simplicio, I ask you a difficult question: Do you think this event is real in the sense in which you use this term?

SIMPLICIO   Of course it is real![1]

SALVIATI   You then think that its existence and reality are quite independent of whether we look at it or not?

SIMPLICIO   Most certainly.

SALVIATI   Or even of whether we record it?

SIMPLICIO   There is no doubt about it.

SALVIATI   Suppose now that point 2 is separated from point 1 by a wall and that to the right of this wall there is an absolute vacuum, so that no recording of the event can be made. Do you think that the event still takes place?

SIMPLICIO   The event is real and objective and therefore quite independent of its observation or recording in any manner whatsoever. Of course, in the absence of any record I could not say exactly at what point the event occurs; I can only make statements of probabilities with the available theory of quantum mechanics, but I think this is merely a sign of the incompleteness of our present theory. I believe that in a more complete theory, still to be achieved, better prediction might be possible.

SAGREDO   Simplicio no doubt thinks that this is not essentially different from the coin-tossing experiment, in which we know for certain, even before we look at the results, that one of two possibilities is realized, since merely looking at the coin has no influence on the outcome of the toss.

SALVIATI   Of course, both of you must admit that this is mere supposition and that there is no possibility of ever verifying or falsifying an entirely different theory, which would say, for instance, that just before the act of looking at the result, the coin reverses itself, head becomes tail, and vice versa.

SIMPLICIO   I believe you are right insofar as the logical possibility is concerned, but you would have a hard time reconciling such a peculiar theory with the laws of mechanics.

SALVIATI   To tell you the truth, I have no intention of seriously proposing such a theory. All I want to establish is that it is neither logically nor phenomenologically refutable.

SIMPLICIO   I suppose I have to grant that.

SALVIATI   So that if there were in the Λ^0 decay experiment phenomenological evidence that the event which we observe in the bubble chamber does not occur in a vacuum, then we would have to accept this as evidence against the "reality," in your sense, of the event.

SIMPLICIO   Can you produce such evidence?

SALVIATI   I cannot do it in a simple manner for this case, but I have arranged a thought experiment which illustrates a closely related situation, in which such a test *is* possible.

I mount a radioactive atom on a fixed point in space opposite a detecting screen covered with photographic emulsion, which records the decay product. After the atom has decayed in the direction of the screen, one of the grains in the emulsion will be blackened. As we repeat the experiment under similar conditions, other grains will also be blackened in a random manner.

Now again, we ask the question: If we remove the screen

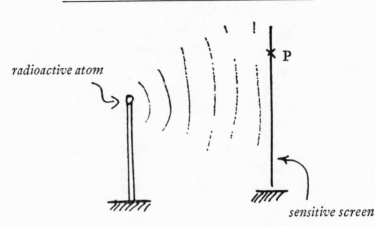

FIG. 6. *Radioactive decay recorded on a sensitive screen.*

or the emulsion, will the individual decay particle still arrive at a certain point when it has reached the distance of the screen?

SIMPLICIO   Since the arrival of a particle at a certain point on the screen is an objectively given fact, which we can record or not, we must conclude that this arrival at some point will also occur if no screen is placed to observe it. Of course in the absence of a screen, such a statement cannot be verified by physical observation, but this is not much different from any other statement of this kind.

For example, if I look at this beautiful tree in your garden, I know that the tree is there, and this knowledge can be confirmed by a host of other correlated observations: I can go there and touch it, I can enjoy the shade it gives me from the sun, etc., etc. But if I turn my back to the tree and do not look at it any more, I still know that it is there, although under these conditions I have no direct means of verifying such a statement by observation. We make such extensions of the world of experience into the world of pure theory all the time; it is the basis of a coherent life. Even in death, we presuppose such an interpretation since life insurance would make no sense at all without such a belief. Such an embedding of the experiences in a theoretical substratum is absolutely essential to give meaning to our life.

Thus when I say that even in the absence of a screen the particle will still be at a certain point at its former position on the screen, I extend an observation into the realm of the unobserved, which seems to me perfectly consistent with common sense.

SAGREDO Simplicio has expressed very clearly what everyone does more or less, even when one is not conscious of how fragmentary is the information on which we construct our coherent image of the world. Nevertheless, it is perhaps good to remember from time to time *how* fragmentary this information is, and that there is absolutely no logical necessity to fill in the vast empty spaces between the phenomena in one particular way or in any way. Thus if we do fill them in with something, we supplement the phenomena with a theoretical and conceptual frame into which we can organize these phenomena in a more or less coherent way.

It is not surprising that the English word *fact* comes from the latin "facere," which means to *make*, as if with that root one wanted to express that facts are not entirely given to us from the outside world, but that they are also made, shaped, and endowed with meaning and significance by the man-made conceptual reference frame. It is just as though we were saying, "What do you *make* of it?"

But I see that you have that distant look on your face, Salviati, from which I always recognize that behind your patience you hide a knowledge which you are preparing to reveal to us. Pray do not spare your words and let us know what you think.

SALVIATI We have indeed arrived at a very crucial

point in our analysis, and what you have just said, Sagredo, concerning our natural tendency to fill the empty spaces between phenomena with theoretical constructs is a warning to us that this natural process, which we carry out at almost every moment of our lives, is not necessarily transferable to the atomic level.

And indeed I shall show that we have solid evidence, based on equally incontestable facts, that Simplicio's answer to my question concerning the actual position of a particle emitted from a radioactive atom cannot be maintained.

SIMPLICIO   This I would like to see.

SALVIATI   It is easy to show! To this end, I have made a slight modification of the experiment. As before, I place a radioactive atom behind a screen, but I have pierced two holes in the screen, and behind this arrangement I have placed a second screen, which is coated with a sensitive surface to record the arrival of individual atoms.

With this experimental arrangement I set up two series of experiments. First I cover one of the holes and then the other, and record the arrival of the particles in the form of the distribution curves C_1 and C_2, respectively. They are just

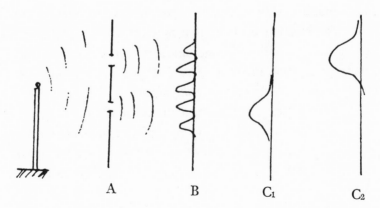

A   B C_1 C_2

FIG. 7. *Transmission of decay particles through two holes.*

what one would expect, and resemble the distribution of bullets on a target sheet.

The important second part of the experiment is now carried out with both holes open. Now if Simplicio is right in his assertion that events occur independently of the observations, one should expect that the particles sometimes pass through one and sometimes the other hole, so that the resultant distribution would simply be the sum of the distributions found on C_1 and on C_2. However, the actual result is quite different. The distribution resembles an optical interference pattern, as indicated on screen B.

Thus Simplicio's answer is contradicted by the experiment.

SIMPLICIO I am aware that people have often discussed such experiments, but I doubt whether they are sufficient to disprove the realist's position as to the objective validity of individual events.

It is clear that the appearance of interference patterns on the screen B would indicate that wave propagation is intimately associated with the propagation of the particles, as, for example, a singularity in the solution of a partial differential equation may propagate itself inside the regular part of the solution. The wave then acts as a sort of guide, which directs the particles to various places in proportion to the intensity of the wave. This does not exclude the possibility that each individual particle actually passes through one of the holes, although they arrive at B with a probability distribution as observed.

This theory, developed by a very eminent French physicist,[2] is in perfect agreement with the facts which you have mentioned, and is at the same time in agreement with the realistic interpretation of the propagation of particles in space.

SALVIATI   It is a very ingenious solution to the problems which one has to face when one wants to maintain a
"realistic" interpretation of these phenomena.

It also shows how entirely different ideological motiva-
tions may actually have much more in common than their
adherents would be willing to admit in public. At least, you
do not seem to be the least embarrassed to accept help for
your threatened faith, wherever it comes from, be it from
the left or the right. But my admiration for this theory is
genuine; it should be taken seriously and studied carefully.

Let us see whether it can be maintained. To this end
let me further modify the experiment. Instead of closing one
hole as I did before, I place into one hole a sensitive counter
which will not interact with the particle more than is necessary to record its passage through this hole. What does your
theory predict for the outcome of the experiment? Shall we
still see the interference pattern or not?

SIMPLICIO   I cannot tell you the answer immediately
because I do not know this theory well enough. You must
give me some time to think.

SALVIATI   Take all the time you need. While you
think, I shall ask Sagredo for his opinion.

SAGREDO   I have become very suspicious of your
questions, Salviati. But if you insist on my opinion, let me
try to see what I can contribute to analyzing the situation.
So long as we talked only of particles it was simple enough.
The answer would have been that there is no change in the
result, since the mere *observation* of the passing particle
would certainly be insufficient to modify the fact that the
particle passed through the hole or did not. If observation influenced the actual velocity distribution of the particle that
passes through the hole, it would be easy enough to determine this influence by closing the other hole and running

two series of experiments with only one hole open, one with the counter in place and one without. After opening the second hole one would thus expect merely a distribution curve on the screen, which would be obtained by adding two curves C_1 and C_2, where one of the two might have to be slightly modified before adding them because of the presence of the counter in one of the holes.

What confuses me now is that this very reasoning would be all the more valid in the *absence* of the counter, and that this is the case when we observe an entirely different result, namely the interference pattern which we observed before on B.

But when we start talking about waves, I believe the situation changes considerably. A wave, in order to produce interference, must be so delicately attuned that the phase relations in the two holes are preserved until the arrival of the wave on the screen. If I place a counter into one hole it may very well be that the phase relations are destroyed and the interference pattern disappears.

Now in Simplicio's theory, there are both particles and waves, and therefore I can obtain an unambiguous conclusion only if I attribute more importance to one of the two descriptions. Since I have no sufficient reason for doing that, I cannot reach a verdict.

SIMPLICIO   I think Sagredo is right. We have to take an empiricist's attitude and adapt our theory to the empirical facts as revealed by experiments of increasing refinement.

SALVIATI   I believe I told you before that a theory which can be adapted to all as yet unknown future facts is no theory at all because it has no predictive power.

SAGREDO   But do not keep us in suspense any longer, Salviati; tell us what the experiment would reveal if we did make the observation with the counter in one of the holes?

SALVIATI I shall be glad to do so. The interference pattern disappears under these conditions, and we observe the distribution corresponding to the sum of the two curves C_1 and C_2 on the screen.

SAGREDO This is a most remarkable result! Is it not as though the observation of one aspect, namely the passage through the hole, has completely wiped out another aspect of the system, namely the interference from the two holes. So that here again we encounter another incident of that complementarity which we have seen before.

It seems as though the very condition which is necessary for the establishment of one of these aspects is incompatible with the disposition needed for revealing the other. Thus complementarity here is not something which is due to the insufficiency of the measuring equipment, but is rather an essential consequence of the very conditions under which an experiment is possible.

SALVIATI The result is all the more remarkable insofar as it is incompatible with Simplicio's theory, no matter how he tries to adapt it.

SIMPLICIO How do you come to this conclusion?

SALVIATI I am sure you are already acquainted with the answer, although you perhaps do not remember it at the moment. Let me help you to remember it by some questions.

By placing the counter into one of the two holes, I find that it counts a particle whenever the particle passes through this hole. What do you think is the counting rate of such a counter compared to the total rate of particles passing through any of the two holes?

SIMPLICIO I think one should expect about half the number of particles which pass through either of the two holes to trigger the counter. The other half pass through the other hole without triggering it.

SALVIATI   Right. And which of these particles will interact with the counter?

SIMPLICIO   Those which trigger the counter will interact with it, of course!

SALVIATI   And the others?

SIMPLICIO   The others will not interact since they pass through the other hole, and if the counter works correctly, they are too far away to have any effect on the counter.

SALVIATI   Right again. Now for those which do not interact, what would you expect for their distribution on the screen?

SIMPLICIO   Since the particles do not interact at all, one should expect a distribution with an interference pattern.

SALVIATI   And for the others?

SIMPLICIO   For the others one should presumably expect the distribution to correspond to one of the curves on C_1 or C_2, since all the particles pass through the same hole, as if the other hole were closed to them.

SALVIATI   For the total effect you would thus obtain the sum of an interference curve with one of the curves on C_1 or C_2.

SIMPLICIO   Wait a minute, not so fast. This would indeed be the conclusion if the counter interacted only with the *particles* in the pilot wave. But if they also interact with the *wave*, then the effect could be different since the presence of the counter could destroy the coherence of the wave passing through the two holes, and thus wipe out the interference pattern.

SALVIATI   I was wondering whether you would see

this loophole to escape the inevitable conclusion which disagrees with the facts. Since you have seen it, I have to ask you a further question: How could you tell whether a counter has interacted with the wave?

SIMPLICIO   By the fact that the interference is destroyed.

SALVIATI   Would such an interaction have any effect on the counter?

SIMPLICIO   According to the universal law of action and reaction it should have such an effect.

SALVIATI   So that sometimes the counter should be expected to count, because of its interaction with the wave?

SIMPLICIO   Yes, I think so.

SALVIATI   Even though the particle itself passes through the other hole?

SIMPLICIO   I guess there is no escape from this conclusion.

SALVIATI   So that finally you can save appearances only by assuming that the counter does something quite different from what it was designed to do. It will sometimes get triggered even if no particle is in its neighborhood.

SIMPLICIO   This seems to be the conclusion, unless I have made a mistake.

SALVIATI   You have made no mistake. Your conclusion is about as reasonable as one can expect from such a theory as you have outlined it to us.

SAGREDO   Salviati, I think you are absolutely incredible. Your thought experiment has completely demolished the beautiful pilot wave theory with the particle sitting inside it like a singularity reaching its destination by the combined action of the wave and external forces.

What a beautiful picture this would have been. I was imagining the particle to be like a surfboard rider bobbing up and down, urged hither and yon by the action of the waves along a beach. But now this picture is blurred since I am no longer able to distinguish the particle from the wave and the whole hole business has become foggy again. It is really a pity!

SIMPLICIO   I doubt very much whether the experiment which you have described, Salviati, can actually be performed with the precision needed to come to such conclusions. It seems to me that such an experiment would be extremely difficult, and I have never heard of its actually being performed.

What you have described to us is only a *thought experiment*, and your results are of course those answers which you would obtain from the application of quantum mechanics. But since it is this very theory which is under investigation here, we should not accept one of its conclusions to prove that another theory which is more complete and realistic cannot be right.[3]

SAGREDO   I think Simplicio is right with his objection, and a thought experiment will not suffice to prove the inadequacy of his theory with the pilot wave. But it seems to me that an experiment has actually been carried out that verifies Salviati's statements, although under somewhat different circumstances.

SIMPLICIO   I surely would like to hear about such an experiment. Can you tell us more?

SAGREDO   Gladly, as far as I can remember it. The idea for this experiment came from Schrödinger, and it was carried out by the Hungarian physicist L. Janossy about five or six years ago. It is a standard optical interference experiment using a Michelson-type interferometer so designed that it is possible to experiment with individual photons.

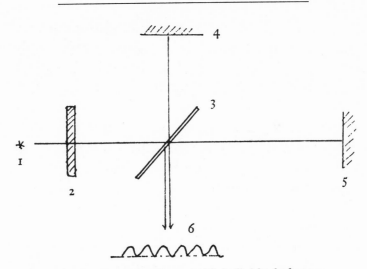

FIG. 8. *Interference experiment with individual photons.*

The light from a source 1 falls on an absorber 2 to reduce its intensity, so that at any given moment only one photon is in the apparatus. The photon then falls on the half-silvered mirror 3, which reflects half the intensity to mirror 4 and transmits the other half to mirror 5.

After having been reflected at these two mirrors, the light arrives again at mirror 3, and half of each intensity appears at point 6, where the two beams are brought to interference.

One can now make two complementary experiments, just as we did with the thought experiment with the two slits. In the first, one can observe the arrival of individual photons at 4 and 5 by replacing the mirror with a photomultiplier tube. This experiment shows that the photons arrive in a random manner at 4 and 5 with a probability exactly proportional to the intensity of the two beams 3–4 and 3–5.

In the complementary experiment one observes the interference pattern at 6 and finds that the position of this pattern depends on the relative distance of *either* of the two mirrors from point 3.

The conclusion of the first experiment is that each individual photon arrives at one of the mirrors independently of the others. The conclusion of the second one is that each individual photon is somehow simultaneously present at each of the mirrors. These conclusions are contradictory and, therefore, cannot both be right at the same time.

Salviati Thus neither of them can be right.

Sagredo And yet they are based on experiments and not on theoretical construction.

Salviati That is not quite so. You will notice that your experimental results, being based on immediate facts, are not contradictory by themselves; they *become* so only if you take these facts as evidence for the existence of a *real* property which is supposed to be present independent of any observation and whose existence is only *revealed* by the observation.

Sagredo We should thus call them complementary properties, which exist only in relation to a detecting instrument.

Salviati Yes, and two such complementary properties can be revealed only by mutually exclusive physical arrangements, where the very presence of one precludes the application of the other.

Sagredo And this is not due to our lack of ingenuity, but it is the very essence of the physical law that we are trying to discover.

Salviati This is indeed so, and once this basic point is recognized, the enigmas of quantum mechanics begin to resolve themselves into a beautiful and consistent theory.

Sagredo A theory, however, which upsets our deep-rooted epistemological prejudices to such an extent that it may have repercussions in many other domains of human knowledge.

SIMPLICIO You two seem to have found a harmonious concurrence, and you seem to be overjoyed at the prospect of living in a world where the reality of properties becomes blurred around the edges. I for one can only register the facts, if indeed they are facts, as they are reported to me, but I cannot get any joy out of this baffling situation.

You said before that this experiment is similar to the two-slit experiment, but I believe this similarity is not complete. In the two-slit experiment we recorded the presence of a particle with a counter which left the particle untouched except for a possible slight deviation in its momentum. In the photon experiment the use of photomultiplier tubes in the first part of the experiment absorbed the photons entirely, so that we can never verify whether the interference pattern is actually destroyed, or whether it is destroyed in a trivial manner simply because no more photons are present to be observed.

SAGREDO What you say is true, Simplicio, and I wonder whether it would not be possible to modify the experiment so as to preserve the photons while still somehow recording their presence. Salviati, do you know whether this can be done?

SALVIATI Yes, it can be done. It was not done by Janossy simply because it is technically much more difficult to do and the experiment was difficult enough without this complication. One can register the presence of a photon by means of an effect which played an important role at the beginning of quantum theory. This is the Compton-effect, or rather an adaptation of it to the case of our mirrors. Since it has not actually been done, we can discuss it only as a thought experiment.

Suppose that we suspend one of the mirrors and measure its momentum before and after the photon has passed through the instrument. Then we can conclude from the conservation of momentum that the photon has been at this mirror if and only if this momentum has changed by twice the momentum carried by the photon, which is $h\nu/c$.

SIMPLICIO   Why twice?

SALVIATI   Because this is the amount of momentum transferred by the photon to the mirror at reflection, when it reverses its direction of motion.

Thus, if the momentum of the mirror before impact is p_1 and after impact is p_2, we find

$$2h\nu/c = p_1 - p_2.$$

Of course, in order to make such a measurement we must be sure that the momentum of the mirror can be measured with sufficient accuracy, so that we have a situation as indicated approximately in our sketch. Thus, we should have for the uncertainty Δp of the momentum

$$\Delta p < 2h\nu/c. \tag{1}$$

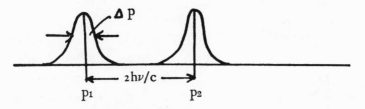

FIG. 9. *Distribution of momentum before and after the collision of a photon with the mirror.*

Of course, if we determine the momentum of the mirror with such accuracy its position is uncertain by an amount of at least

$$\Delta x = h/\Delta p, \tag{2}$$

so that because of (1) we find

$$\Delta x > \tfrac{1}{2}\, c/\nu = \tfrac{1}{2}\, \lambda$$

where λ is the wave length of the light.

But if Δx is uncertain by such an amount then the phase of the reflected photon is also uncertain by the amount π, and this means, of course, that an interference becomes impossible.

SAGREDO This is a beautiful analysis, and it shows clearly that the complementarity which we find here is an essential consequence of the quantal properties of all of nature, no matter how large may be the objects under investigation.

But it also shows another aspect. In order to understand the behavior of quantal systems when they produce events, the time evolution of the states of a closed individual system does not contain sufficient information as to the actual occurrence of such events. We must study the evolution of coupled systems which interact for a while and subsequently are separated again.

Do you agree with me, Salviati?

SALVIATI Yes, I do, and in fact I would go further and say that the analysis of the time evolution of interacting quantal systems is by far the most important task which needs to be done before we can hope to understand the occurrence of events in quantal systems.

SIMPLICIO Isn't the evolution of the states of any system determined by an equation known as *Schrödinger's equation*, and isn't this equation a first-order differential equation with respect to the time variable, so that the Schrödinger function evolves as completely causal?

SAGREDO I think you want to say "deterministic."

SIMPLICIO Yes, that is what I meant to say. And further, the equation is symmetrical with respect to the past and the future, so that its solution in both directions of time is completely determined by the value of the state vector at any given instant, for instance, the present moment.

SAGREDO   That is, indeed, the case.

SIMPLICIO   Is this not the most convincing demonstration that the evolution of the Schrödinger state vector as we teach it to all physics students cannot be in agreement with the facts?

SAGREDO   Why not?

SIMPLICIO   Because it cannot describe any events, since events are facts, real facts I mean, which happen out there in the world; and facts which occur only with certain probabilities cannot occur in accordance with a deterministic evolution in both directions of time.[4]

SAGREDO   This is indeed true and it is in agreement with Salviati's contention that the Λ^0 does not really decay suddenly as we see it on the bubble chamber picture as long as we let the state of the Λ^0 evolve in accordance with the Schrödinger equation.

SIMPLICIO   This confirms my views that the Schrödinger state vector does not describe the actual state of an individual system, but merely furnishes the physicist with a kind of calculation tool with which he can make probability calculations for observable events on an ensemble of identically prepared systems.

SAGREDO   I must admit that you have a very puzzling situation here, and I cannot immediately see what I could bring up against this view. What is your opinion about this, Salviati?

SALVIATI   Simplicio's argument is based on a hypothesis which is made almost universally by physicists, including some of my best friends. It is taught in almost all the textbooks. Yet it is wrong. The argument is essentially the following: The state of a quantal system is represented by a vector ψ_t in a unitary space depending on the time t in such a manner that the evolution can be described by a continuous unitary group V_t so that $\psi_t = V_t \psi$.

From this incorrect hypothesis Simplicio's incorrect conclusion follows by his perfectly correct reasoning.

The above hypothesis is, however, true only if the system under investigation has no interaction with the exterior, so that it is what we call a *closed* system.

If the system does interact with another one so that it becomes an *open* system, then this description of the evolution is incorrect. Every such evolution is partly a *stochastic* evolution and is not reversible in time.

The very fact that you observe the occurrence of events presupposes precisely that the system under observation interacts with another one. Thus, under these conditions the evolution is that of an open system and for that system it is not deterministic and not reversible.

SIMPLICIO   I do not understand a word of what you are saying. Are we discussing quantum theory as it is taught to us in the schools and in hundreds of textbooks, or are we discussing your own private theory? Let us make up our minds! I think your own private theory is of little interest to us at the moment since quantum mechanics is the theory which is confirmed by a large number of experiments in all parts of the world. I have never heard your theory with the stochastic evolution of a system that interacts with another one.

SALVIATI   Calm yourself, Simplicio. It is not my fault that you do not recognize quantum theory when I explain it to you. Perhaps you have never really understood it, because what I am saying is in complete agreement with the conventional theory although it is not always emphasized as much as it should be. And not only that, but the stochastic evolutions of a quantal system are seen by hundreds of physicists every day.

SAGREDO   Aren't the relaxation phenomena of paramagnetic resonance experiments of this type when a spin in a pure state interacts with other spins or with the lattice vibrations and gradually evolves from a pure state into a statistical mixture?

SALVIATI   A very good example indeed. And in fact every radioactive decay registered by a counter or any other detecting system is an equally good example.

SAGREDO   Now, I understand for the first time, I believe, the paradox of Schrödinger's cat, which has caused me much concern, especially because I recently lost my beautiful cat Annabelle in a sad accident. There was no question of her being quite dead, and that was a fact that shook me up. The thought that there might be a state in which the cat is a superposition of being alive and of being dead is so preposterous that I can feel only sympathy with anyone who loses faith in the soundness of quantum theory when he hears about it for the first time. If there is no alternate answer to Schrödinger's paradox, nobody who loves animals can possibly believe in quantum mechanics as it is taught us.[5]

SIMPLICIO   Hey, not so fast. If you say you have understood Schrödinger's cat paradox I can only admire you, because you do better than most people. I suspect, however, that you have not really understood it, because to my simple mind there seems to be an unbridgeable gap between the states of the entire "system," atom, counter, hellish contraption, cat, and all, and the state of the cat which is either dead or alive after the experiment.

SAGREDO   You are not so simple minded as you want us to believe. The distinction between the entire system as

FIG. 10. *Schrödinger's cat. The animal trapped in a room together with a Geiger counter, and a hammer, which, upon discharge of the counter, smashes a flask of prussic acid. The counter contains a trace of radioactive material—just enough that in one hour there is a 50% chance that one of the nuclei will decay and therefore an equal chance the cat will be poisoned. At the end of the hour the total wave function for the system will have a form in which the living cat and the dead cat are mixed in equal portions.**

a whole and the subsystem of the cat plus hellish contraption is so minute that it always slips away with our unconscious attitude of ignoring the atomic system from which the entire macroscopic process originated. Since the difference between the two states that Schrödinger and all of us are worrying about is equally small, we are always losing sight of it just at the moment when we need it to explain the paradox.

Are we not facing something here that is very similar to another paradox which we observe when we want to explain in terms of classical mechanics the irreversible behavior of thermodynamic systems in the approach to equilibrium? The classical equations of motion, no matter how complicated the system may appear, are strictly reversible in time while the

*Picture and explanation taken from Bryce S. DeWitt, "Quantum Mechanics and Reality," *Physics Today*, September 1970, p. 30.

thermodynamic systems are not. Yet a physical object should not change its symmetries simply because we visualize it once as a mechanical system and once as a thermodynamic one.

The solution of this paradox was found to lie in the occurrence of fluctuations which are always present but which are generally ignored for thermodynamic systems. This is perfectly sufficient, however, since with the exception of some singular points such fluctuations are usually negligible. As soon as the theory of fluctuations is included in the description of thermodynamic systems, the symmetry between the past and the future is restored and the paradox disappears.

It seems to me, therefore, very plausible that whenever we discuss the quantum theory of two coupled systems we must always keep track of which system we are actually discussing. If it is truly a subsystem of the entire system then the state of such a subsystem, although related in a definite way to the entire system, is not one which evolves according to the usual Schrödinger equation. Instead, as Salviati has just explained, it is a stochastic evolution. Each individual system, in an assembly of similarly prepared ones, will choose a different road of evolution restricted only by a definite law of probabilities.

Here, too, we encounter a most significant example of complementary behavior, exemplified in this case by the pair of complementary properties: causal evolution of a joint system on the one hand, and stochastic evolution of a partial subsystem on the other.

SIMPLICIO If you are right, then this phenomenon should already be present on the microscopic level. Imagine two spins, each of value $\frac{1}{2}$ interacting with each other, one belonging to one particle and one to another. After they have interacted they may be in a state of the form

$$\Psi = \alpha \; u_1 u_2 + \beta \, u_1 v_2 + \gamma \, v_1 u_2 + \delta \, v_1 v_2, \qquad (1)$$

where α, β, γ, and δ are four complex numbers satisfying the equation $|\alpha|^2 + |\beta|^2 + |\gamma|^2 + |\delta|^2 = 1$, and where u_1 represents the state "spin up" of particle 1 and v_1 represents the state "spin down" of particle 1, and similarly for the other two variables, u_2, v_2 of particle number 2.

Now if we consider particle 1 separately and independently of particle 2 then it should be in a state of its own, quite different from state (1). What would this state be?

SAGREDO   The best way to describe such a state would be as a statistical mixture of two orthogonal states, say, with respective probabilities p and q ($p + q = 1, p \geqq 0$, $q \geqq 0$). If we denote the projections onto these two vectors by P_x and P_y then the state could be concisely represented by the von Neumann density matrix

$$W_1 = p \, P_x + q \, P_y. \qquad (2)$$

For particle 1 this is a 2 by 2 matrix given by

$$W_1 = \begin{pmatrix} |\alpha|^2 + |\beta|^2, & \alpha\gamma^* + \beta\delta^* \\ \alpha^*\gamma + \beta^*\delta, & |\gamma|^2 + |\delta|^2 \end{pmatrix} \qquad (3)$$

and the values of p and q are

$$p = \tfrac{1}{2} \, (1 + \sqrt{1 - 4D})$$
$$q = \tfrac{1}{2} \, (1 - \sqrt{1 - 4D}) \qquad (4)$$

where D is given by

$$D = |\beta\gamma - \alpha\delta|^2.$$

I will let you calculate the two state vectors x and y, since it is a simple exercise in algebra. A similar calculation gives the state w_2 for the second system.

SIMPLICIO   How you can figure all this out so quickly is a mystery to me. Actually the exact description of the state of one of the subsystems is of little importance to me; what matters is which of the two descriptions represents the *real* state of the system?

SAGREDO   We are always coming up against the same problem at the crucial points of our discussions. Here we are again facing an impossible duality if we insist on endowing the physical system with a degree of *reality* as we are accustomed to doing in classical physics.

Just as in the two-slit experiment, where neither of the two descriptions was more real than the other, here, too, we have to adapt our concept of reality to the physical situation under which this concept reveals itself.

If we carry out experiments pertaining to the joint system, then no doubt state (1) is the correct expression. If, on the other hand, we carry out experiments pertaining only to system 1, then state (2) is correct. There is never any chance of conflict between the two because the two kinds of experiments are actually incompatible, and therefore cannot be executed on one and the same system.

SIMPLICIO   But if one of the subsystems is my measuring apparatus and the other is the system which I measure with it, then your description of the situation implies that there are in principle still measurements possible on the joint system with a second measuring device which would determine properties describable only with a state (1). This seems to me incompatible with the occurence of *objective events* in the first measuring apparatus.

SAGREDO   Yes, you are quite right. At this point we must define more precisely what it means to produce an *objective event* in a measuring apparatus. But I prefer to call on Salviati to comment on this since he understands this point much better than I, and he has not said anything for a long time.

SALVIATI   I was fascinated by the dialectical process which you two were following in your dispute, and I thank you both for elucidating several points along the way. Although I have pursued this road of reasoning so many times that I often feel I have exhausted the subject, when I followed your progress just now I was again deeply impressed by the significance of the epistemological revolution that quantum theory has forced upon us.

You left off at just that point where the analysis of the measuring process must be completed by an analysis of the meaning of an *unambiguous* and *objective event.*

The necessity of understanding the meaning of objectivity and unambiguity is all the more urgent since complementarity, which we have encountered so far in various circumstances, seems to have washed out this concept. Yet everyone would agree that science is impossible without the firm anchor of objectivity which alone will enable us to discourse profitably on scientific matters. This is so obvious that it hardly needs repetition.

Since all our information on the physical universe is ultimately based on the experience of phenomena associated with observation and experimentation, the assurance of the objective character of such observations is a matter of primary concern for us.

When we try to analyze these properties with sufficient precision we are led to the following considerations:

The result of a measurement should serve as a basis of a scientific theory. Therefore it should be *unambiguous* in the sense that the alternatives which are singled out by the various possible values of the observable quantities should all be compatible. Furthermore the measurement should reveal an *objective* property in the sense that a repetition of the measurement under the same conditions by the same or any other observer would yield the same result.

Observations which satisfy these criteria can in general be carried out only by physical apparata large enough to behave with a very high degree of probability like classical systems. This means that whatever their ever-present quantal properties may be, they are in the context of measurement of negligible importance and therefore inessential to the measuring process. In order to establish permanent records for such observations of a macroscopic kind it is also necessary that the system exhibit certain ergodic properties which guarantee that the quantal processes serve as triggers for the macroscopic amplification processes.

SAGREDO   Now I understand why all atomic measuring instruments are always equipped with a kind of amplifier that starts from a highly unstable equilibrium state and evolves in a cascade process toward a final state. This is certainly so for counters of all kinds, including photomultipliers, bubble and Wilson chambers, Geiger counters, and photographic emulsions.

SALVIATI   Yes, all these devices serve to amplify an event to the level of a datum that can be perceived by human beings without further instruments and without interference with the essential content of the record.

However, the amplifying device itself should not be considered as the essential part of the measuring device since it serves merely to render perceptible an event that has already occurred. Thus, the essentially anthropocentric character of this stage is no serious objection since it is not the most important part of the measuring apparatus, although it is certainly indispensable from a practical point of view.

SAGREDO   If this is so, may we not try to formulate axiomatically the essential properties of a measuring instrument by saying that such an instrument must permit the choice between classical alternatives and no others. This is the exact translation of your definition of *unambiguous* and of *objective*, and means, in the formalism of quantum mechanics, that the only observable quantities which can be measured with a particular apparatus are represented in this formalism by commuting projection operators.

SALVIATI   I believe that such a characterization of the measuring instrument is possible, but I must express a warning concerning the absolute precision of such an axiomatic formulation. The word "classical" cannot single out such clear-cut properties as you imply with your axiomatic representation. It should be used only in an approximate sense. Indeed, any concrete model of the measuring process based on the essentially quantal nature of all material bodies shows only that the nonclassical features will be absent with an overwhelming probability in the sense of statistical physics. The axiomatic treatment of the measuring process would thus seem to idealize the situation more than it need be.

SAGREDO   I understand exactly what you mean and I am convinced that you are right on this point. But in spite of this, it seemed to me of interest to bring out the ideal classical properties of a measuring apparatus with full precision and clarity which can be furnished only by the axiomatic characterization of such a system. The situation is similar to those which we have noticed several times, when an essential point can be made more easily with idealized elements and concepts in a theory. For instance, this is the case when we use irrational numbers. Strictly speaking, such numbers have no counterpart in reality, yet they are extremely useful.

Of course, I am no doubt presumptuous in trying to contradict you in a matter which you understand so much better than I could ever hope to.

SALVIATI I think your point of view offers no serious inconvenience so long as you keep in mind the limitation of your method. In particular, in such a process one must always remember that no formal rendering can ever be complete, that it must leave undefined a certain number of "primitive" concepts and take for granted, without further analysis, certain relationships between these concepts. The concrete meaning of these primitive concepts and axioms can only be formulated in a "meta language" which is outside the formal scheme of the theory. The particular difficulty in this situation is that the proper choice of these undefined concepts and of their interpretation presupposes a complete knowledge of the entire physical situation. Thus at the outset there is a mutual interdependence of the physical content and its conceptualization which can never result from a simple, logical process.

Once one has understood this point one is necessarily more reluctant to attach a great deal of importance to the axiomatic-deductive organization of a theory such as quantum mechanics, although on a superficial level or for didactic purposes such a procedure might be quite useful.

SAGREDO What you say is certainly most significant and lends credence to the belief that there is much more in science than the mere observation and recording of events and their integration into a conceptual structure.

There are also vision and creative imagination, qualities which alone enable us to abstract from the multitude of possible phenomena those which reveal the true nature of reality. Science reveals structures which are significant or meaningful in some sense and the more meaningful they are, the more real they are.

Is this not the answer to the unresolved task in Simplicio's dream? Are they not the ingredients needed to bring to light the beauty and perfection of creation, all hidden away in Simplicio's library, but inaccessible to him because he did not pay attention to what is essential in the search for truth?

Among those principles, that of *complementarity* is no doubt the sum and substance of our experience with the phenomena of microphysics. Instead of being a principle which expresses the limitation of our ability to know, it expresses the very essence of the objective rendering of the physical phenomena in the unambiguous language pertaining to factual evidence. Once its general character is recognized its operation is seen in many areas where objectivation of experience takes place. In particular, the behavior of quantal systems furnishes us with new points of view concerning the essential properties of composite systems, where it is recognized that the sum of the parts does not at all exhaust the properties of the whole.

SALVIATI I do agree, Sagredo, and I have on several occasions formulated similar ideas. I would go further and add that the understanding of the behavior of individuals interacting within a group is incomprehensible without new structural and dynamical categories which cannot be derived from individual behavior.

Just as the correlations between interacting quantal systems lead to many different kinds of mutations in the behavior of individual systems, which, for want of a better term, we call quantum jumps, so does the person integrated in a group produce spontaneous insights which would have been inaccessible to him in isolation.

Thus—and I address these words particularly to Simplicio—does our science of microphysics lead to insights which transcend the domain from which they originated, insights which give reasons for hope of a better understanding of all our experiences, including the moral and social behavior of man.

SAGREDO I believe that this is perhaps a fitting moment to conclude our dialogue for today, since your words, dear Salviati, are so filled with meaning that anything we might say after them would seem shallow.

There remains for me the pleasant duty of thanking you both for your contributions to our discourse, from which all of us have certainly profited. Who knows, a fortunate occasion may bring us together again at some other time. I sincerely hope so. And so I bid you farewell until we meet again.

Notes

First Day

1. The reference is to the *Mémoire*: "Sur la double réfraction que les rayons lumineux éprouvent en traversant les aiguilles de cristal de roche, suivant les directions parallèles à l'axe." It contains the first correct description and interpretation of the phenomena of the polarization of light.

2. Simplicio identifies himself as a relativist here. He definitely rejects the ether as a carrier of the vibrations of light, a notion which dominated the thinking of the last half of the nineteenth century. The crucial experiment is that of Michelson and Morley, which failed to detect any "ether wind" due to the movement of the earth. It was the starting point of the theory of relativity developed by Einstein in 1905.

3. The original Simplicio in Galileo's dialogue was identified with the philosophy of Aristotle, which, in the seventeenth century, played a role comparable to that of dialectical materialism today. Sagredo, by reminding Simplicio of his switch, uses this occasion to underline the fact that our scientific notions are not independent of our ideologies.

4. This is the first of several traps. Particles and fields are complementary concepts. A unified theory containing both entities at the same time requires the renunciation of attributes of reality, which Simplicio is not prepared to do. His answer shows that he is not aware of the pitfalls, and he naively affirms his belief in the "reality" of these concepts.

5. This is the point where the complementary aspect of particles and fields is exploited to challenge Simplicio.

6. This comment by Sagredo is no doubt only partially sincere. It plays too well into Salviati's hands.

7. This remark by Salviati points up the essential weakness of all hidden variable theories: If hidden variables are needed to render atomic events causal, why should we stop at that point and not admit all kinds of occult causal relationships.

8. This is an allusion to Thomas S. Kuhn's perceptive essay *The Structure of Scientific Revolutions* (Chicago: University of Chicago Press, 1962), where great emphasis is placed on the discontinuous character of scientific progress.

9. The reference here is to Lenin's work on *Materialism and Empirocriticism*.

10. Plato's dialogue *Timaeus* was the only work of the Greek philosophers which was available in Latin during the early Middle Ages. It had an inordinate influence on medieval thinking, and gave a completely distorted picture of Greek philosophy, with which its content was often identified.

11. See note 5 of the Second Day.

12. The first proof of the irrationality of $\sqrt{2}$ is found in Euclid (*Elements* X, Appendix 27), but tradition attributes its discovery to the Pythagoreans, who considered it a mystery and imposed secrecy about it on the members of the Pythagorean community. It is a proof by "reductio ad absurdum," and was mentioned by Aristotle as a typical example of this kind of proof (*Prior Analytics* I.23). It is short and elegant:

Suppose $\sqrt{2} = m/n$, where m and n have no common divisor. It follows that

$$2n^2 = m^2,$$

which shows that m is even; say $m = 2p$. Thus

$$2n^2 = 4p^2, \text{ or } n^2 = 2p^2.$$

This shows that n is even, too; $n = 2q$. Hence m and n are both even, and thus they have 2 as a common divisor. This is a contradiction with the hypothesis. Hence $\sqrt{2}$ is irrational; q.e.d.

This discovery caused great embarrassment to the Pythagoreans since it seemed to conflict with their fundamental conviction that every measure could eventually be reduced to integers.

In spite of its obvious failure in number theory, Pythagoreanism, in various forms, has played a very important role in the history of science down to the present, and some of the most important discoveries were made under its influence.

13. In the psychology of C. G. Jung the "archetypes" are the contents of the collective unconscious. These are symbolic images appearing spontaneously at certain stages in the development of the individual, either in dreams, fantasies, or artistic and literary productions. Salviati expresses here the opinion that these "irrational" symbols are also responsible for some of the scientific concepts which appeared spontaneously in the history of science.

An example of such a connection is the recent discovery of higher symmetries in elementary particle physics. There is a conscious (or unconscious) allusion to ancient Buddhist mysteries in the review of these theories entitled *The Eightfold Way*, by M. Gell-Mann and Y. Ne'eman (Menlo Park, Cal.: W. A. Benjamin, 1965). Such higher symmetries are in fact closely related to the unifying archetype of the "mandala" found so prominently displayed in Hindu art. Even the representations of these symmetries with the root and weight diagrams have more than a superficial relation to such pictorial symbols.

Second Day

1. Salviati makes an allusion here to the fact that there are actually a number of very different theories which are denoted as hidden variable theories. The simplest and most natural is the one that attributes to every individual in an assembly of identically prepared systems a number of variables which determine with certainty the outcome of any possible measurement on that individual system.

Such hidden variables are now no longer considered seriously by the experts since it could be established that it is impossible to construct statistical ensembles with such variables in accord with the known properties of such ensembles (J. M. Jauch and C. Piron, *Helv. Phys. Acta*, 37:293, 1964).

A weaker form of hidden variables are those for which some of the properties which are normally associated with propositions of an individual system are relaxed. Thus, if for a physical system proposition a is true and another proposition b is true, then one would normally think that the proposition a *and* b is true too. If one drops this assumption then one obtains a more general class of hidden variables among which the so-called local hidden variables lead to some interesting conclusions which deviate from ordinary quantum mechanics (J. S. Bell, *Physics*, 1: 195, 1965).

These conclusions were recently tested by two series of experiments, one using the correlation of photon polarization in the annihilation of positronium (L. Kasday, Thesis, Columbia University, 1970) the other using the correlation of polarization of two photons emitted in a cascade (J. F. Clauser, M. A. Horne, A. Shimony, and R. A. Holt, *Phys. Rev. Lett.*, 23: 880, 1969).

The latest and decisive experiment was carried out by S. J. Freedman and J. F. Clauser, *Phys. Rev. Lett.*, 28: 938 (1972).

A third form of hidden variables consists in introducing variables for modifying the time evolution of states so as to simulate the random sequences that we observe in quantal experiments. Such a theory was recently developed (D. Bohm and J. Bub, *Rev. Mod. Phys.*, 38: 453, 1966), and it was also subject to an experimental test (C. Papaliolios, *Phys. Rev. Lett.*, 18: 622, 1967). The result is negative too. There are a number of other ideas for such theories, which have not yet been subjected to any tests.

2. The reference here is to A. Einstein.

3. This is a classic example, which had already been used by Hume in his famous discussion on causality.

4. The reference is to Max Born, who often discussed this point, especially in the later years of his life. The example of the wheel is a slight variation of an example invented by Einstein. He considered a point moving on a straight line between two perfectly reflecting walls.

5. The "Academician" was the way Galileo refered to himself in the original dialogue. He was the proud member of the Academia dei Lincei, founded by Prince Cesi in Rome, whose main objective was the establishment of the Copernican system.

6. Galileo was often accused of dealing with idealized situations only, which could never be realized in the actual world. This was one of the greatest assets of Galilean physics and one of the reasons for its successes. It was also a danger; and in the case of his theory of tides, for example, it led him completely astray.

7. This is a reference to Galileo's fascination with circular movement. Even after Kepler, with enormous labor and against all tradition, had established the elliptical shape of the planet Mars's orbit, Galileo continued to persist in his belief in circular orbits. He

could never make the switch from Archimedes to Apollonius, a strange weakness in a scientist as bold and original as Galileo.

Sagredo a bit mysteriously alludes to the possibility that circular movement might be significant in a much more profound sense than Galileo could ever have known. His remark can be interpreted to mean that the quantum condition for periodic motion can be meaningfully extended to general motion through the Fourier integral theorem.

8. Sagredo does indulge here in a little bit of sophistry by sliding almost imperceptibly from the preceding discussion on general laws and individual boundary conditions to the very different epistemological problem of the uniqueness of the individual and the unscientific nature of the unique.

THIRD DAY

1. The semireligious setting, the dim light, and the mystery are all symbols which underscore the ideological involvement in Simplicio's psychic state.

2. Ideologies are in a sense also social adaptations, and the better they function the less is required of the individual to differentiate himself from the others in the crowd. The process of individuation begins with the awareness of the "other," who usually seems to the beholder much better adapted than himself.

3. This is it! The fatal word has been pronounced. It is the source of Simplicio's anxiety, and it represents the existential counterpart of determinism.

4. The "wise old man" is a well-known archetype discovered by C. G. Jung in countless dream sequences. The encounter with the "wise old man" means that the resources of collective racial wisdom are at the disposal of the dreamer if he learns to read and assimilate the symbolic message of the dream.

5. The luminous stone is the "lapis philosophorum" of the alchemists. It symbolizes the integration of the personality.

6. The Maxwell demon was invented by Maxwell in his famous discussion on the irreversibility of thermal systems. It is an imaginary creature (or device) which can control the movements of individual atoms. The demons cannot actually function in physical systems

because they are subject to the same fluctuations as the atoms which they are to control.

The fact that they can function in the dream means that they are creatures from another level of reality than the dreamer, and thus carry the message of the existence of deeper levels of consciousness which are essential for the psychic processes about to be initiated in Simplicio.

7. The mysterious woman whispering into Simplicio's ear is the archetype of the "anima." She is the harbinger of instinctive truths yet to be learned by Simplicio. The passage from the number three to four symbolizes a fundamental problem in individual psychology that is related to the acceptance and unification of opposites. In the context of the dialogue it represents the symbolic acceptance of the principle of complementarity, for which Simplicio is not yet ready.

8. The self-assured intellectual male looks foolish next to the instinctive female, who, with one simple gesture, can upset his entire value system.

9. This is the heart of the dream's message: What is the good of winning the whole world if one loses one's own soul? Evidently there are two ways of winning and losing. The first is the one to which the anima was referring just before the game started. The other is the one which we now see occurring. Simplicio is not yet capable of distinguishing the two, hence his surprise at "winning" when just a moment before he believed the anima that he was going to lose.

10. The gold represents the material gain and also, on a symbolic level, the objective of alchemy.

11. The color green is the symbol of hope and of a new start. It often occurs in dream sequences at a decisive moment in the individual's history, when options leading to new prospectives become available.

12. This is typical of the materialist's psychic disposition. He believes that there is no problem which cannot be solved by more or bigger material things.

13. The idea of the "complete" library has been discussed many times before. It occurred to the author from reading "The Library of Babel" by J. L. Borges, in *Labyrinths* (New York: New Directions, 1962). It poses an interesting paradox. For instance, being complete the library must also contain all its catalogues, including the catalogue of all the catalogues and so on. Evidently such a library

cannot be finite. This problem is closely related to Russell's paradox.

14. We know that the letter referred to here was actually written by Galileo since the replies from his daughter in the convent at Arcetri are preserved. The letters from Galileo to his daughter are unfortunately lost. They would be the most important source of information on Galileo's trial, since in them he gives a complete account of everything that happened to him when he was questioned by the inquisition in Rome.

15. This is the point where the naked truth is revealed to Simplicio. The ultimate wisdom is not to be found in quantity. There is the other side, so far completely overlooked by Simplicio. There are two criteria of truth as Einstein told us and, as the dream shows, the neglect of the second one leads to absurdity. Cf. A. Einstein, Autobiographical notes, in P.A. Schilpp, ed., *Albert Einstein, Philosopher-Scientist* (Evanston, Ill.: Library of Living Philosophers, 1949), p. 21.

16. Evidently Simplicio is not a positivist; neither is Salviati, however. No positivist would have spoken as he did at the end of the first day. But they reject positivism for different reasons. Simplicio's reason is the *reality* of physical properties, which positivism rejects or at least puts in doubt. Salviati's rejection is for *methodological* reasons. Phenomena for him are not a sufficient basis for constructing a theory. This theme will be developed further during the third day. Simplicio thinks his accusing Salviati of secretly being a positivist is the supreme insult that he could throw at Salviati, a natural feeling for a dialectical materialist.

17. This story of Salviati's friend is borrowed from a humorous poem by Christian Morgenstern, with the punch line:

> Eingehüllt in feuchte Tücher
> studiert er die Gesetzesbücher . . .
> und so schliesst er messerscharf
> dass nicht sein kann, was nicht sein darf.

18. Simplicio refers here to Galileo's (alias Salviati's) theory of the tides, which is no doubt Galileo's greatest scientific blunder and a most remarkable example of wishful thinking. This theory was supposed to furnish him the proof of the Copernican system. It was the only proof he had, and he stuck with it like the captain of a sinking ship.

19. This reference is to A. Einstein.

20. This passage was written by J. A. Wheeler, one of the foremost experts on this theory.

Notes

Fourth Day

1. Simplicio again falls into the trap. He does it so resoundingly that there seems to be no hope of his seeing the light.

2. The reference is to Louis de Broglie, discoverer of matter waves by a theoretical argument, for which he was awarded the Nobel prize.

3. Simplicio's last-ditch defense.

4. Simplicio does not express himself very clearly here but what he wants to say is essentially the following: Suppose that one of a series of input events $A_1, A_2, \ldots$ occurring for a physical system at a time $-T$ produces at the time $+T$ a series of output events $B_1, B_2, \ldots$ with certain probabilities $p_1, p_2, \ldots$. A reversal of the time evolution of such a system starting with a particular one of the events $B_1, B_2, \ldots$ at time $+T$ will *not* result in one of the events $A_1, A_2, \ldots$ at time $-T$ with certainty, but only with a probability distribution $q_1, q_2, \ldots$, say. Thus, the symmetry between past and future is destroyed as soon as events have occurred.

5. Sagredo refers here to the well-known paradox of Schrödinger's cat, which we reproduce here, for the benefit of the readers who might not be familiar with it, Schrödinger's own words, translated from the German (cf. *Naturwiss.* 23: 807, 1935):

> A cat is placed in a steel chamber, together with the following hellish contraption (which must be protected against direct interference by the cat): A Geiger counter contains a tiny amount of radioactive substance, so tiny that within an hour one of the atoms may decay, but it is equally probable that none will decay. If one decays the counter will trigger, and via a relay activate a little hammer which will break a container of cyanide. If at the end of an hour the cat is still living one would say that no atom has decayed. An indication of the first decay would be the presence of equal parts of the living and the dead cat.
>
> The typical feature in these cases is that indeterminacy is transferred from the atomic to the crude macroscopic level, which then can be *decided* by direct observation. This prevents us from accepting a "blurred model" too naively as a picture of reality. By itself it is not at all unclear or contradictory. There is a difference between a blurred or poorly focused photograph and a picture of clouds or patches of fog.

THE TEXT OF THIS BOOK was set on the Linotype in a face called JANSON, a robust "Old Face" of the Dutch school. Having in mind the proclivity of the early punch-cutters to wander, it is interesting to note that this type was cut in Amsterdam by a Hungarian named Nicholas Kis, *circa* 1690. It was erroneously named for the Dutchman Anton Janson, who had been employed in Leipzig, where the original matrices were discovered years later. These same mats are today in the possession of the Stempel Foundry, Frankfurt, and the version you are reading was modelled directly on type produced from the original strikes.

The typography is by GUY FLEMING, and the composition by HERITAGE PRINTERS.